ESSENTIALS OF
FORENSIC IMAGING
A TEXT-ATLAS

ESSENTIALS OF
FORENSIC IMAGING
A TEXT-ATLAS

ANGELA D. LEVY, MD
H. THEODORE HARCKE, MD

CRC Press
Taylor & Francis Group
Boca Raton London New York

CRC Press is an imprint of the
Taylor & Francis Group, an **informa** business
A TAYLOR & FRANCIS BOOK

CRC Press
Taylor & Francis Group
6000 Broken Sound Parkway NW, Suite 300
Boca Raton, FL 33487-2742

First issued in paperback 2020

© 2011 by Taylor and Francis Group, LLC
CRC Press is an imprint of Taylor & Francis Group, an Informa business

No claim to original U.S. Government works

ISBN-13: 978-0-367-57702-5 (pbk)
ISBN-13: 978-1-4200-9111-3 (hbk)

Library of Congress Cataloging-in-Publication Data

Levy, Angela D.
    Essentials of forensic imaging : a text-atlas / Angela D. Levy, H. Theodore Harcke Jr.
      p. ; cm.
    Includes bibliographical references and index.
    Summary: "This groundbreaking work emphasizes a practical, technical, and interpretive foundation for understanding the diagnostic use of modern imaging in forensic autopsy. It covers advanced imaging techniques, such as radiography, multidetector computed tomography (MDCT), angiography, and magnetic resonance imaging (MRI). The authors focus on images derived from MDCT and MRI that are likely to replace or limit conventional autopsy dissection. Following an introduction to basic concepts and a review of imaging techniques, chapters are organized by cause of death. The atlas-like presentation is supported by full text explanations that review relevant forensic principles"--Provided by publisher.
    ISBN 978-1-4200-9111-3 (hardcover : alk. paper)
    1. Autopsy. 2. Diagnostic imaging. 3. Autopsy--Atlases. 4. Diagnostic imaging--Atlases. I. Harcke, H. Theodore. II. Title.
    [DNLM: 1. Autopsy--methods--Atlases. 2. Diagnostic Imaging--methods--Atlases. 3. Forensic Medicine--methods--Atlases. W 617]

RA1063.4.L48 2011
614'.1--dc22

2010029284

Visit the Taylor & Francis Web site at
http://www.taylorandfrancis.com

and the CRC Press Web site at
http://www.crcpress.com

To our nation's Fallen Warriors who have passed through the Charles C. Carson Port Mortuary at Dover Air Force Base, Delaware, on the way to their families and final resting places. It is through the investigation of their deaths that we have been able to learn new things that will advance forensic medical practice and lead to changes that make service members safer in the future. We have endeavored to serve them all, at all times, with dignity, honor, and respect.

CONTENTS

PREFACE

The concept for this book evolved from our experience with teaching radiologic pathologic correlation and the unique collaboration between the Department of Radiologic Pathology and the Office of the Armed Forces Medical Examiner at the Armed Forces Institute of Pathology. The core theory of radiologic pathologic correlation is that an understanding of the pathologic basis of disease is fundamental to the interpretation of radiologic images. This book brings the long tradition of radiologic pathologic correlation to forensic radiology and autopsy.

We are fortunate in our practice to have the ability to interpret multidetector computed tomography (MDCT) images just before the autopsy and receive immediate feedback from experienced forensic pathologists. When we started interpreting postmortem MDCTs, we tried to read the images without knowledge of the history and external physical findings. We then moved to the autopsy table where we could directly correlate imaging findings with autopsy findings. This process helped us retain objectivity as to the strengths and weaknesses of postmortem MDCT, and it taught us how we could best add value to the autopsy process. The dialogue that takes place at the autopsy table about the imaging and autopsy findings and the discussion about the possible mechanism of injury and cause of death is the ultimate radiologic pathologic correlation experience. We work with an experienced team of talented professionals that includes forensic pathologists, anthropologists, dentists, DNA experts, investigators, and photographers. This rich experience provided a foundation for us to develop a broad knowledge base in forensic imaging.

In writing this textbook, our goal is to provide the reader with a technical and interpretive foundation for understanding the application of modern cross-sectional imaging, namely MDCT, to forensic autopsy. To this end, we integrated imaging material with basic forensic information in a topical format. As an atlas, the emphasis in this book is placed on the rich pictorial display of case material demonstrating the power of MDCT imaging. The text provides classic radiologic pathologic correlation by concise discussions of the autopsy and radiologic findings such that the reader can learn the diagnostic value of imaging applied to forensic autopsy. The autopsy photographs provide a striking complement to the images. The resulting radiologic pathologic correlation provides a unique and informative instructional tool suited to those oriented in either imaging or forensics.

Forensic imaging with MDCT and other modalities is a rapidly growing and changing field. We highlighted the strengths and limitations of advanced imaging in the major categories of death seen in our practice: gunshot wounds, blunt trauma, blast injury, thermal injury, drowning, natural deaths, sharp force injury, suicide, and a few less common causes such as asphyxiation and electrocution. We do not attempt to comprehensively cover clinical forensic medicine or the entire breadth of forensic autopsy because there are excellent textbooks available. However, these textbooks currently lack imaging material. We believe this book will be useful to forensic pathologists, radiologists, and investigators, particularly those who have added or are considering the addition of MDCT to their practice. It is our expectation that some readers will become the leaders in the new subspecialty of Forensic Radiopathology.

Angela D. Levy, MD
H. Theodore Harcke, Jr., MD

ACKNOWLEDGMENT

The authors appreciate the assistance of the Armed Forces Medical Examiner Captain Craig T. Mallak, Medical Corps, U.S. Navy, in the preparation of this text. His expertise as a forensic pathologist and attorney contributed to the clarity of our discussions and helped strengthen the radiologic pathologic correlation that is essential to this manuscript.

THE AUTHORS

Angela D. Levy, MD, earned her medical degree from the University of Texas Southwestern Medical School and is certified by the American Board of Radiology. She served as chairman of the Department of Radiologic Pathology and chief of Gastrointestinal Radiology at the Armed Forces Institute of Pathology. As an associate professor of radiology at the Uniformed Services University of the Health Sciences (Bethesda, Maryland), she was chief of Abdominal Imaging. She retired from the U.S. Army Medical Corps as a colonel and is currently a professor in the Department of Radiology at Georgetown University Hospital, Washington, D.C.

H. Theodore Harcke, Jr., MD, FACR, FAIUM, earned his medical degree from Pennsylvania State University College of Medicine and is certified by the American Board of Radiology. He is a graduate of West Point and is currently a U.S. Army Medical Corps colonel serving as chief of Forensic Radiology in the Department of Radiologic Pathology at the Armed Forces Institute of Pathology, Washington, D.C. He holds appointments as professor of radiology and pediatrics at Jefferson Medical College and adjunct professor of radiology at the Uniformed Services University of the Health Sciences.

Chapter 1

Introduction to Forensic Imaging

Technological advances in the last decade led to revolutionary changes in cross-sectional imaging. Developments in multidetector computed tomography (MDCT) and magnetic resonance imaging (MRI) technology transformed the collection and display of image data. The image resolution of MDCT and MRI scans increased as the acquisition times decreased. Computer workstations and multiplanar two- and three-dimensional software enable radiologists to view and interpret images in ways that were not previously possible. As a consequence of these technological advances, cross-sectional imaging fundamentally changed the practice of clinical medicine such that noninvasive or minimally invasive diagnostic and therapeutic techniques are commonplace.

The application of these technologies to forensic medicine is a natural extension of clinical imaging. As MDCT and MRI scanners are being incorporated into forensic facilities worldwide, the benefits and limitations of these technologies in forensic investigation are still being explored (Hayakawa et al. 2006, Poulsen and Simonsen 2007, Rutty et al. 2008). Undoubtedly, incorporation of cross-sectional imaging techniques in forensic medicine represents a new approach to forensic radiology and affords an opportunity to increase the contributions of radiology to forensic medicine and add objective, reproducible anatomic data to autopsy.

Our experience with postmortem MDCT evolved from the September 11, 2001, terrorist attack. The department of Radiologic Pathology at the Armed Forces Institute of Pathology supported the Office of the Armed Forces Medical Examiner in the forensic investigation of the victims of the Pentagon attack. The conventional objectives of forensic radiology were applied to the investigation—human remains identification and assistance in the forensic assessment of the crime scene. As the investigation proceeded, the efforts of radiology focused primarily on remains identification, because the recovered bodies were subjected to extreme physical and thermal trauma, and in many cases, human remains were mixed with debris from the site (Figure 1.1). Radiography was used to establish whether human remains, personal effects, or other materials were present in recovered specimens (Harcke et al. 2002). Some of the lessons learned from this experience were carried forward in the design of a new mortuary facility, the Charles C. Carson Port Mortuary at Dover Air Force Base, Delaware, which opened in 2003. As part of a research initiative, MDCT was installed in the mortuary in 2004.

After the first week of scanning, it was clearly evident that MDCT would become an invaluable part of the autopsy process and it was quickly incorporated into all autopsies performed at the mortuary. The images we obtained during the first week of scanning showed us that postmortem MDCT had tremendous potential. MDCT produced images with exquisite anatomic detail that displayed fracture patterns that were difficult to assess at dissection (Figures 1.2 and 1.3) and multiplanar and three-dimensional full-body images of ballistic and skeletal trauma (Figure 1.4). In some cases, we were able to assess gunshot wound tracks and injury comparable to autopsy (Figures 1.5 and 1.6). We recognized that there were limitations in the assessment of visceral and vascular injury because the technique did not use intravascular contrast media, and MDCT three-dimensional surface rendering could not match the external examination of an experienced forensic pathologist (Figure 1.7 through Figure 1.9).

Since our initial experience with these victims of ballistic trauma, we have had the opportunity to image a comprehensive range of causes of death and compared them to the complete autopsy that followed. We learned that MDCT provides a wealth of anatomic data prior to autopsy that are invaluable in many but not all causes of death. We also continue to obtain radiographs in all cases. Radiography is considered to be essential because of its excellent resolution and absence of artifacts. It is also the imaging modality of choice to evaluate subtle bone detail and dissociated body fragments and for the anthropologic evaluation of skeletal remains. In contrast to radiography, the role of MDCT in our practice is to provide a two-dimensional multiplanar and three-dimensional anatomic survey prior

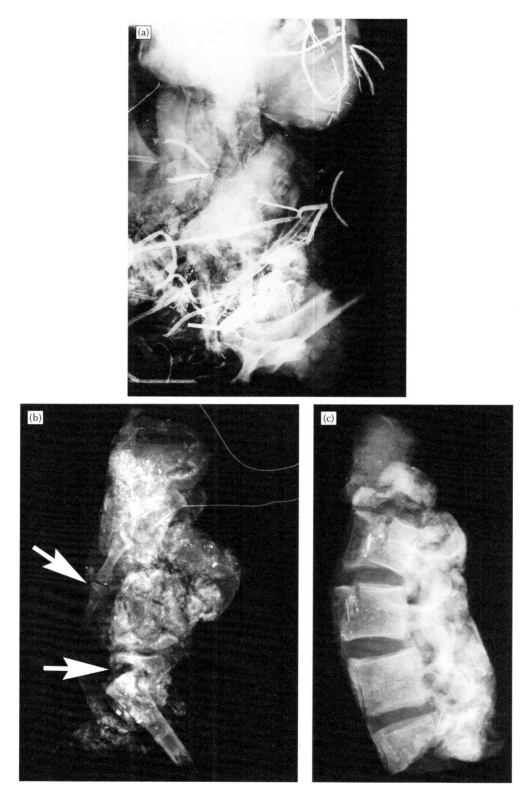

FIGURE 1.1   Radiographs from the investigation of the September 11, 2001, terrorist attack on the Pentagon were obtained to assist in the recovery and identification of human remains. (a) Multiple bone fragments are present and commingled with wires and natural debris. (b) Debris containing bone fragments of skeletally immature victim. Arrows show physeal plates of a finger and wrist. (c) Fragment of a victim's spine shows embedded debris and vertebral body fractures.

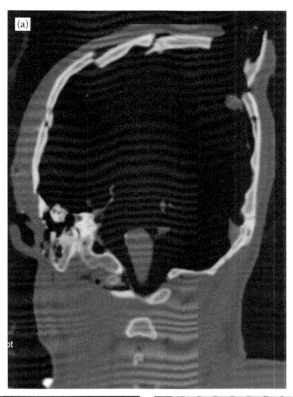

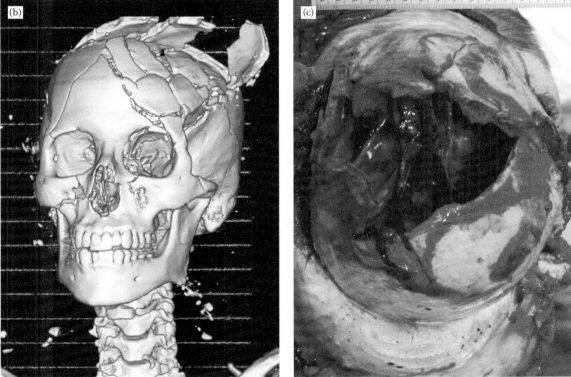

FIGURE 1.2 Lethal gunshot wound to the head. (a) Coronal MDCT shows an air-filled cranium and complex, comminuted skull fractures created by a bullet that traveled through the skull. (b) Three-dimensional MDCT shows the position and pattern of the fracture fragments with an intact scalp. (c) Autopsy photograph of the skull after the scalp has been removed shows the fracture and injury. The fracture fragments dissociate from one another when the scalp is removed, limiting the appreciation of the fracture pattern.

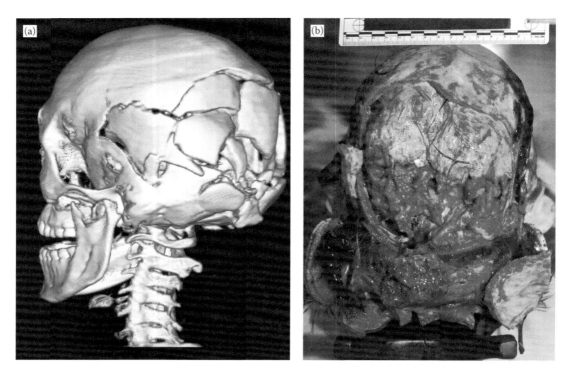

FIGURE 1.3   Lethal gunshot wound to the head. (a) Three-dimensional MDCT shows a complex, comminuted posterior skull fracture. (b) Autopsy photograph of the cranial dissection shows that the fracture fragments fall apart when the scalp is rolled away from the skull making it difficult to appreciate the orientation and relationship of the fracture fragments to one another.

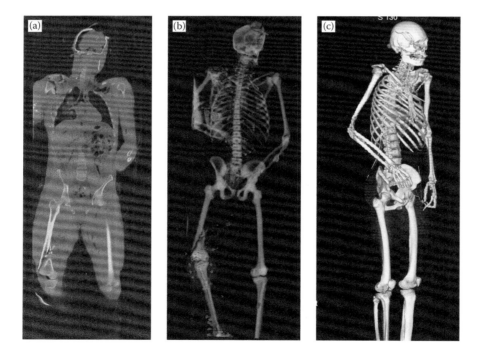

FIGURE 1.4   Full-body MDCT reconstructions. (a, b) Victim of multiple gunshot wounds to the head, torso, and extremities. Full-body coronal MDCT in (a) shows fractures of the skull, right shoulder, and femur. There are bilateral pneumothoracies and pulmonary injury and retained metallic fragments in the right chest and pelvis. Three-dimensional MDCT in (b) shows blue color tagging of metallic fragments, revealing the distribution of metallic fragments throughout the body. (c) Victim of blunt trauma to the head shows a right temporal skull fracture on a three-dimensional MDCT.

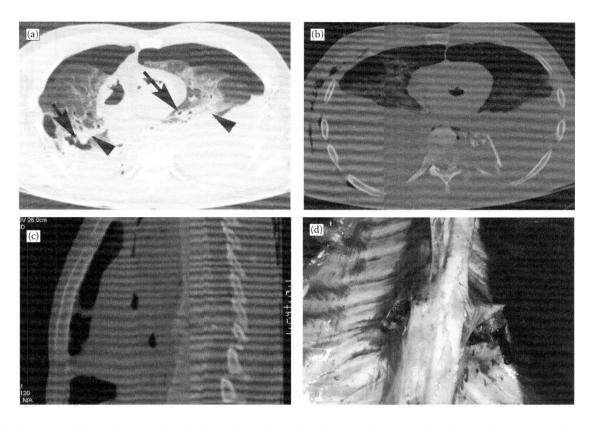

FIGURE 1.5 Lethal gunshot wound to the chest. (a) Axial MDCT in lung windows shows large bilateral hemopneumothoraxes and the gunshot wound path marked by gas (arrows) and increased attenuation from hemorrhage (arrowheads). Rib fractures and spinal fractures are also present. (b, c) Axial and sagittal MDCT in bone windows shows a complex fracture of the thoracic spine with displaced vertebral body fracture fragments that extend into the spinal canal and left chest. (d) Autopsy photograph of the anterior aspect of the thoracic spine shows the entry and exit wounds through the thoracic spine. Further dissection revealed complete transection of the spinal cord that was not evident on MDCT.

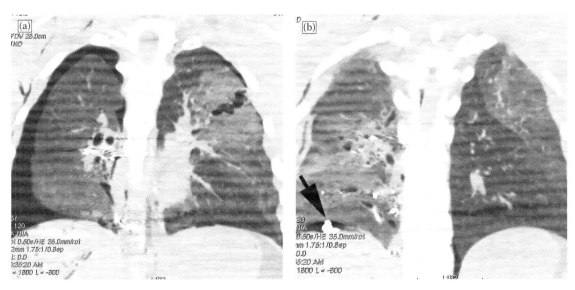

FIGURE 1.6 Lethal gunshot wound to the chest. (a, b) Coronal MDCT in lung windows shows the bullet track through the lungs. Gas, hemorrhage, and metallic fragments in the bullet track delineate its path. The more anterior section (a) shows the gunshot wound track in the left upper lobe, and the more posterior section (b) demonstrates the gunshot wound track in the right lung. Bullet fragments are present in the right lower lung and adjacent to the right hemidiaphragm (arrow).

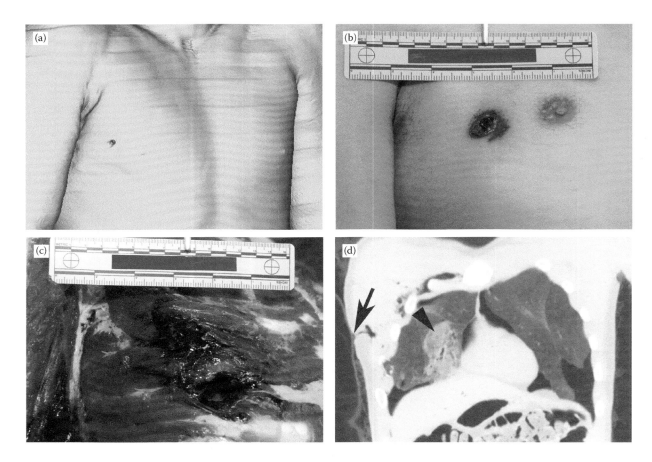

FIGURE 1.7   Lethal gunshot wound to the chest. (a) Three-dimensional skin surface rendering shows the gunshot entry wound to the right of the right nipple. (b) Autopsy photograph shows the entry wound with marginal abrasion and soot. These findings are not evident on the surface-rendered image. (c) Autopsy photograph of the chest dissection shows hemorrhage surrounding the wound track in the right ribs. (d) Coronal MDCT shows the entry wound in the chest wall (arrow) containing gas and gas in the chest wall. There is focal hemorrhage in the lung (arrowhead) in the wound track and bilateral pneumothoraxes.

to dissection that guides the forensic pathologist to specific abnormalities if necessary. MDCT is also used to evaluate and document complex fracture patterns of the skull, spine, and pelvis which are often difficult to assess at autopsy. Postmortem MDCT adds objective and reproducible anatomic data to forensic autopsy.

Interpretation of image findings in the postmortem state is uniquely different from clinical image interpretation. One of the most important considerations when adapting MDCT and other cross-sectional imaging modalities to forensic medicine is that an understanding of the postmortem state and principles of forensic autopsy is necessary for accurate interpretation of the images.

PRINCIPLES OF FORENSIC AUTOPSY

Forensic autopsy is an autopsy performed for the medicolegal investigation of death. In the United States,

approximately 20% of deaths undergo medicolegal investigation annually (Godwin 2005). A forensic pathologist performs forensic autopsies under the provisions of local, state, or federal law through a coroner or medical examiner system. The indications for forensic autopsy include sudden or unexpected death; death associated with criminal violence; an unexpected or unexplained death in an infant or child; death associated with police action; death of an individual in local, state, or federal custody; death caused by workplace injury; apparent electrocution; death associated with alcohol, drugs, or poison; unwitnessed or suspected drowning; unidentified bodies; skeletonized or charred bodies; and when the forensic pathologist deems it is necessary to determine the cause and manner of death or collect evidence (Peterson and Clark 2006).

The purpose of medicolegal death investigation is to determine the identity of the deceased; establish the cause,

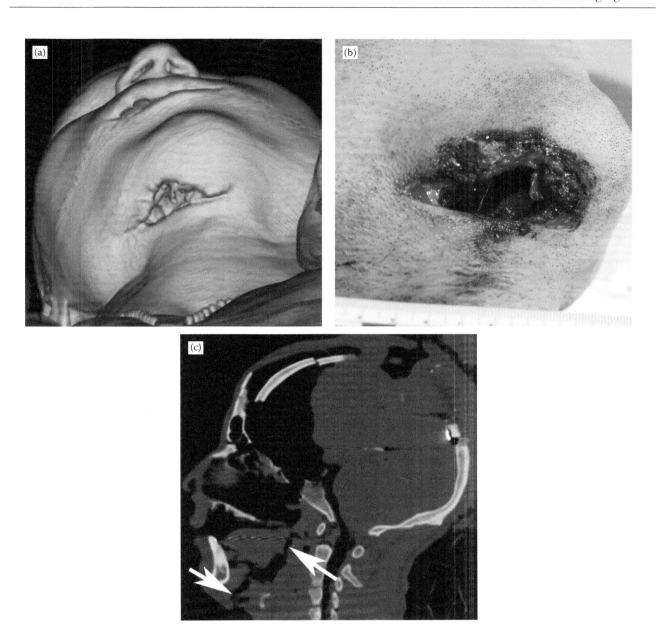

FIGURE 1.8 Self-inflicted gunshot wound. (a) Three-dimensional skin surface rendering of the submental region shows an irregular marginated wound. (b) Autopsy photograph shows the entry wound with marginal abrasion and soot. These are important forensic findings that indicate it is an entry wound and there was contact of the gun muzzle with the skin surface. These findings are not present on the surface-rendered MDCT. (c) Sagittal MDCT shows gas in the wound path in the submental region and through the tongue (arrows). The path is not directly visualized in the brain. There are small foci of gas and metallic fragments in the brain. The exit wound is the large defect in the vertex of the skull. The skull and skull base are extensively fractured.

mechanism, and manner of death; identify contributory factors to death; document injuries and natural disease; collect medicolegal evidence; and in some cases, provide expert testimony. The cause of death is the underlying disease or injury responsible for initiating the sequence of events that led to death. The mechanism of death is the physiologic process that caused death. The manner of death is a description of how the death occurred or the circumstances in which the death occurred. It is the opinion of the forensic pathologist that is derived from the autopsy findings, circumstances surrounding death, and information from the scene investigation. The manner of death is described as natural, accidental, homicide, suicide, or undetermined.

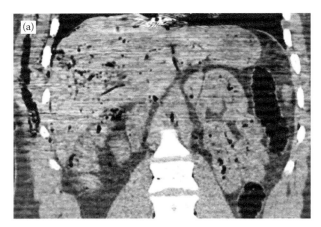
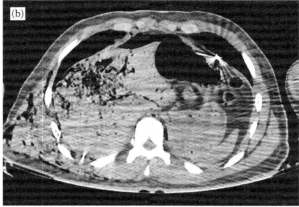

FIGURE 1.9   Gunshot wound to the abdomen. (a, b) Coronal and axial MDCT of the upper abdomen show extensive gas collections in the liver from a gunshot wound. Scattered metallic fragments are present in the posterior right and left anterior abdomen. There is a pneumoperitoneum, pneumothorax, and gas in the subcutaneous tissues of the right flank. The exact path of the bullet cannot be clearly identified. Specific abdominal vascular injury and hemorrhage within the liver cannot be appreciated without intravenous contrast material.

Forensic autopsy is not performed in all medicolegal death investigations. The medical examiner may choose not to perform an autopsy or to perform a limited or complete autopsy. Ancillary and supportive procedures such as radiography, laboratory tests, histology, and toxicology are also performed at the discretion of the medical examiner. In the United States, the National Association of Medical Examiners (NAME) provides performance standards to serve as guidelines for the minimum standards for each component of a forensic autopsy (Peterson and Clark 2006). NAME recommends that radiography be performed for all infant deaths, gunshot wound and explosion victims, and charred remains.

ROLE OF RADIOLOGY IN FORENSIC AUTOPSY

Immediately upon its discovery in 1895, radiography was used in the courts to document and illustrate gunshot wounds (Brogdon 1998). For many years, the principal role of radiography has been to aid in the determination of identity when conventional methods such as fingerprinting or DNA analysis are not available or cannot be utilized; to assist in the evaluation of injury, namely bone injury in blunt and ballistic injury; and to localize metallic fragments and foreign bodies.

Adding MDCT to forensic autopsy expands the role of radiology in forensic autopsy, allowing the radiologist or

forensic pathologist to view anatomy without dissection. The multiplanar and three-dimensional capability makes the anatomic display of the MDCT similar to that of autopsy (Ljung et al. 2006, Thali et al. 2003). Cross-sectional imaging viewed in this manner directs the forensic pathologist during dissection, allows injury patterns to be visualized in three dimensions, detects occult disease or injury, and enables thorough evaluation of anatomic areas that are difficult to dissect. In certain causes of death and forensic scenarios, it is possible that cross-sectional imaging may serve as a triage technique to help forensic pathologists decide which decedents should have an autopsy or if the autopsy should be limited or complete. In those cases that do not undergo autopsy, cross-sectional imaging findings add anatomic information to the external examination and toxicology findings that were previously used alone to determine the cause of death.

Interpretation of postmortem images requires an understanding of the changes that occur in the body after death, namely postmortem change and decomposition. For example, gas is produced during decomposition, so it is often normal to find intravascular gas. Knowledge of the position of the body at the time of death may be key to resolving imaging findings. It is also important to keep in mind that supine positioning of the body on the scanner table was not necessarily the position of the body at the time of death and that there may have been anatomic shifts in organ position because of an injury that was inflicted to cause death. For example, the location of a stab wound track and the organ position with respect to the stab wound will be affected by whether the victim was sitting or standing at the time of the stabbing. A gunshot wound to the chest may cause the lungs to collapse or pleural space to fill with blood, and this affects the position of the gunshot wound track on postmortem imaging.

ROLE OF IMAGING IN THE DETERMINATION OF CAUSE, MECHANISM, AND MANNER OF DEATH

Incorporation of cross-sectional imaging into forensic medicine expands the role of radiology such that imaging not only provides supplemental and supportive information for the determination of cause and mechanism of death, but may also be the primary source of information for these determinations. For example, a postmortem MDCT of a gunshot wound victim may show a gunshot wound track through the lungs with a large hemothorax. The cause of death is a gunshot wound to the chest. The mechanism of death is fatal hemorrhage. But there are limitations to MDCT. It may not reveal the exact site of vascular injury that produced the hemorrhage, and if bullet fragments were present, autopsy would be necessary to recover them. In contrast, other causes of death such as atherosclerotic cardiovascular disease may have normal postmortem MDCT scans and autopsy would be necessary to determine the cause of death. The strengths and limitations of a particular imaging modality should be considered when applying them to a particular case. The manner of death cannot be determined by cross-sectional imaging or radiography findings alone. Although radiologic findings may support the opinion of the forensic pathologist, the consideration of intent to harm, violence, and circumstances surrounding death requires information from the scene of death and witnesses.

CONCLUSIONS

Radiology is an integral part of forensic autopsy. Radiography alone is used in most centers. However, technological advances in cross-sectional imaging have made it possible for MDCT to be used routinely with forensic autopsy. Cross-sectional imaging makes the radiologic contribution to forensic autopsy more effective and brings the potential to increase both the speed and accuracy of forensic pathologists and anthropologists. Throughout this textbook, we discuss and illustrate the MDCT features of various causes of death with autopsy correlation. The benefits and limitations of imaging are highlighted for each cause of death.

REFERENCES

Brogdon, B. G. 1998. *Forensic radiology,* Boca Raton, FL: CRC Press.

Godwin, T. A. 2005. End of life: natural or unnatural death investigation and certification. *Dis Mon* 51: 218–277.

Harcke, H. T., Bifano, J. A., and Koeller, K. K. 2002. Forensic radiology: response to the Pentagon Attack on September 11, 2001. *Radiology* 223: 7–8.

Hayakawa, M., Yamamoto, S., Motani, H. et al. 2006. Does imaging technology overcome problems of conventional postmortem examination? A trial of computed tomography imaging for postmortem examination. *Int J Legal Med* 120: 24–26.

Ljung, P., Winskog, C., Persson, A., Lundstrom, C., and Ynnerman, A. 2006. Full body virtual autopsies using a state-of-the-art volume rendering pipeline. *IEEE Trans Vis Comput Graph* 12: 869–876.

Peterson, G. F., and Clark, S. C. 2006. Forensic autopsy performance standards. National Association of Medical Examiners Annual Meeting (October 16, 2006), San Antonio, TX.

Poulsen, K., and Simonsen, J. 2007. Computed tomography as routine in connection with medico-legal autopsies. *Forensic Sci Int* 171: 190–197.

Rutty, G. N., Morgan, B., O'Donnell, C., Leth, P. M., and Thali, M. 2008. Forensic institutes across the world place CT or MRI scanners or both into their mortuaries. *J Trauma* 65: 493–494.

Thali, M. J., Yen, K., Schweitzer, W. et al. 2003. Virtopsy, a new imaging horizon in forensic pathology: virtual autopsy by postmortem multislice computed tomography (MSCT) and magnetic resonance imaging (MRI)—a feasibility study. *J Forensic Sci* 48: 386–403.

Chapter 2

Integrating Imaging and Autopsy

The emergence of multidetector computed tomography (MDCT) and magnetic resonance imaging (MRI) as postmortem imaging techniques has brought unique challenges to radiology and forensic medicine. Incorporating cross-sectional imaging modalities that were specifically developed to image living patients in a clinical setting into forensic medicine requires an adaptation of our understanding of the anatomic appearance of living tissues and human physiology to the postmortem state. The purpose of this chapter is to discuss the integration of cross-sectional imaging techniques into the forensic environment; the technical considerations that are necessary for the interpretation of cross-sectional images, namely MDCT; and radiology reporting.

INTEGRATING CROSS-SECTIONAL IMAGING INTO A FORENSIC ENVIRONMENT

Workflow

Although the standard practice in forensic facilities is to radiograph and image human remains before physical autopsy begins, the timing and circumstances of the imaging can be varied based on the workflow in a particular forensic facility. All possible options that comply with prescribed forensic guidelines should be considered when establishing a forensic workflow that includes radiography and cross-sectional imaging. The decision on when to obtain radiographs and cross-sectional imaging should be based on the processing scheme of the facility and the physical location of the radiologic equipment.

At our institution, intake photography and identification precede radiologic examinations. When human remains are received in radiology, they are imaged in the body bag that they arrived in at the mortuary. The body bag is opened prior to radiography and MDCT to move the zipper and metal grommets out of the imaging field of view. All physical evidence, such as clothing, jewelry, or other items on the body, remains in place while obtaining radiographs and MDCT. After the radiologic exams are completed, the remains undergo forensic dental examination followed by

autopsy. Clothing and personal effects are removed from the body in the autopsy room, and more photographs are taken after the body is cleaned. External examination and dissection complete the process. During dissection, photographs, tissue, and fluid samples are obtained as necessary. While this workflow has advantages of efficiency and maintains an organized chain of custody, radiographs and MDCT images show external debris, clothing, and personal effects on the body. In some cases, these can be detractors because they may create artifacts on the images. However, the imaging does reflect the physical relationship of clothing, personal effects, and in some cases, medical devices at the time of death. If there is significant artifact from external debris, clothing, or personal effects, reimaging after the body is cleaned may be necessary.

There are advantages to imaging the body after it has been cleaned and had a preliminary external examination. Some facilities may find that this fits their workflow better, particularly if the imaging equipment is at another location and transport could affect external evidence. Artifacts from clothing and debris are eliminated, and metallic markers can be placed on the skin surface to identify wounds so that their precise location appears on the images. If imaging is performed after external examination, medical devices (e.g., tubes, lines, and catheters) should be kept in place so their position and possibly their effectiveness can be assessed on the images. Consideration should be given to separately radiograph clothing and other personal effects to check for metallic evidence or other findings that may have been overlooked. The disadvantage of this sequence is the disruption of the continuity between external examination and dissection.

Advantages and Limitations

The advantages and limitations of integrating imaging into forensic autopsy in specific causes of death are discussed in remaining chapters of this book. We found that cross-sectional imaging adds objective, reproducible anatomic data to autopsy. The ability to review and reconstruct the anatomic findings at the time of autopsy at a later date is invaluable. When used as an adjunct to autopsy, cross-sectional imaging provides an anatomic overview of the body prior to dissection. Abnormalities are localized more efficiently when the images are reviewed prior to autopsy; anatomic areas that are difficult to access at dissection are evaluated on images; complex fracture patterns are better visualized; and occult injury may be detected.

Cross-sectional imaging may potentially serve to augment a limited autopsy or, in some cases, replace autopsy. Although small published series, case reports, and anecdotal experience suggest that postmortem MDCT and MRI may be very useful tools to exclude occult trauma or disease in a limited autopsy and to triage cases for autopsy in mass casualty scenarios, careful consideration should be given to the decision to exclude dissection. To our knowledge, there are no published studies that verify the accuracy of using cross-sectional imaging alone or in combination with limited autopsy in establishing the cause of death. Ongoing and future study comparing the accuracy of cross-sectional imaging to autopsy is necessary to validate and establish the effectiveness of imaging modalities in the determination of the cause of death.

The most important limitation in integrating cross-sectional imaging with autopsy is the cost and availability of MDCT and MRI scanners and personnel. Purchase and installation of state-of-the-art equipment in a forensic facility may be prohibitive for some jurisdictions. As an alternative to an on-site scanner, medical examiners may choose to collaborate with local radiology practices or hospitals in order to obtain imaging studies on decedents. When making the decision to use postmortem MDCT and MRI, it is important to realize the strengths and weaknesses of each imaging modality, because each modality has unique benefits and limitations.

TECHNICAL CONSIDERATIONS
State-of-the-art radiography, MDCT, MRI, and ultrasound produce large quantities of digital data. Incorporation of these imaging modalities into a forensic facility requires careful consideration of the space, power, and personnel requirements to operate and maintain these technologies. In addition, a computer network and picture archiving and communication system (PACS) network is necessary if digital radiography, MDCT, MRI, or ultrasound are to

be utilized. PACS replaces traditional film and allows efficient storage and rapid transmission of images to computer workstations for interpretation or viewing of images. PACS networks also have the capability of remote access, which enables radiologists and forensic pathologists to remotely view and manipulate images for diagnosis, collaboration, and educational activities. At our institution, digital radiography and MDCT are integrated into a PACS network that enables image viewing and manipulation in the radiologist reading area. A computer monitor is mounted adjacent to each autopsy table so the medical examiners may view images while performing the autopsy (Figure 2.1).

Radiography
Radiography is the most widely used radiology technique in forensic medicine. Radiographs are used to document fractures, injury patterns, and occult injuries; localize foreign bodies and metallic fragments; and aid in the identification of human remains when conventional methods such as fingerprinting or DNA analysis are not available or cannot be utilized (Brogdon 1998, Fatteh and Mann 1969, Mann and Fatteh 1968). Radiography is invaluable in the forensic investigation of gunshot wounds and is universally used to locate the bullet, identify the type of ammunition and weapon used, document the path of the bullet, and assist in the retrieval of the bullet. Radiography is also the imaging modality of choice to evaluate subtle bone detail such as metaphyseal fractures in child abuse and in the anthropologic evaluation of skeletal remains or dissociated body parts. We consider radiography to be essential even when MDCT is available. Because of its excellent resolution and absence of artifact, radiographs complement cross-sectional imaging. Also, imaging of small dissociated parts mixed with debris is more efficient with radiography.

Radiography can be performed with mobile or fixed radiographic units. If mobile units are employed and radiography is performed in the autopsy rooms, radiation protection measures should be strictly enforced to protect all personnel. Mobile units may also be used as backup when fixed units are nonoperational or as the primary unit for isolation or contamination cases and in field or temporary morgues. Fixed radiographic units in dedicated radiography rooms are the most optimal choice in a dedicated forensic facility (Figure 2.1a). State-of-the-art radiography units are

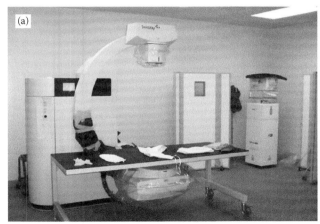
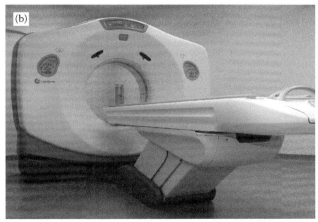
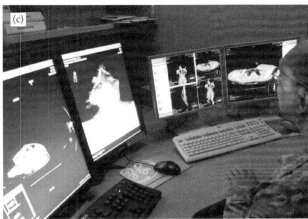
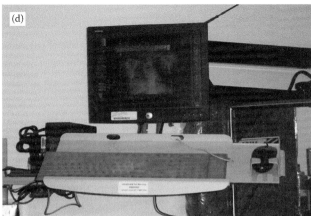

FIGURE 2.1   Forensic imaging equipment used by the Office of the Armed Forces Medical Examiner. (a) Swissray ddrRMulti for digital radiography. (b) GE Lightspeed 16 Xtra MDCT with extended table travel. (c) Radiology reading area with a picture archiving and communication system (PACS) and three-dimensional workstation. (d) Wall-mounted computer monitor adjacent to the autopsy table allows the medical examiners to view images at the time of autopsy.

digital, and the images are viewed on the PACS network. Traditional radiography units require radiographic film screen systems and equipment for wet or dry processing.

Radiographic protocols should be standardized to avoid error. In our institution, all human remains are radiographed in the anterior-posterior (AP) position from head to toe (Figure 2.2). Full-body radiography is necessary even when the suspected wound is in one anatomic location, because additional injury or unsuspected pathology may be found. This is especially true in gunshot wound cases because bullets often travel to unexpected locations in the body. We add a lateral view of the skull to the protocol in all head injury cases (Figure 2.3). If human remains are dissociated, the dissociated parts are cataloged and radiographed separately.

Standard labeling of radiographs designates right and left sides. Labeling should always be done to avoid error. There may be instances when it is difficult to determine the right versus left side of the body. All efforts should be made to determine the correct side of the body. This should be done by the forensic pathologist (medical examiner) and not by radiology personnel. If it is not possible to establish right or left side (e.g., an amputated extremity), a marker should not be placed on the image. By convention, the radiograph is oriented such that the right side of the body is placed on the radiologist's left for viewing (Figure 2.2).

C-Arm Fluoroscopy

C-arm fluoroscopy is most often used to facilitate the localization and recovery of metallic fragments or foreign bodies at autopsy when based upon the localization of the object, on radiographs they are not readily retrievable at

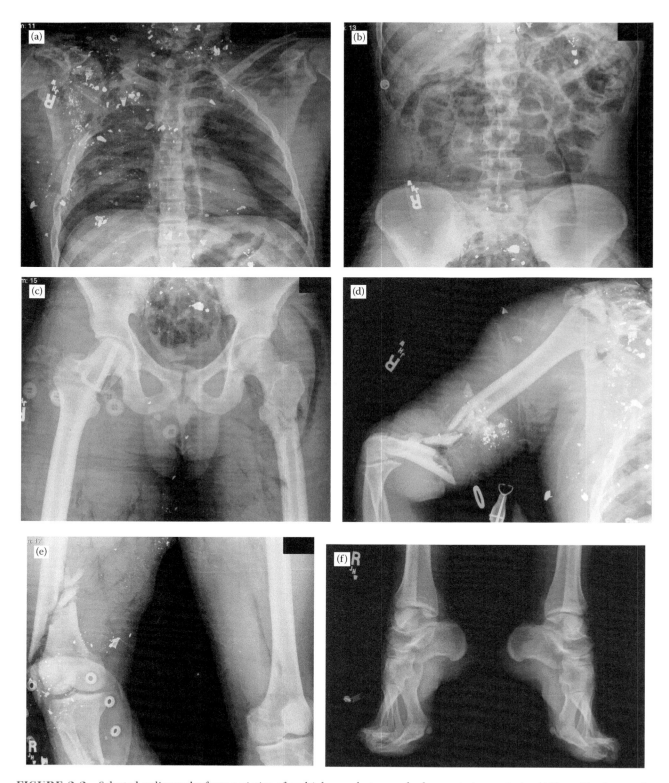

FIGURE 2.2   Selected radiographs from a victim of multiple gunshot wounds show anterior-posterior (AP) positioning on the chest, abdomen, pelvis, and extremity radiographs (a through f) and correct labeling.

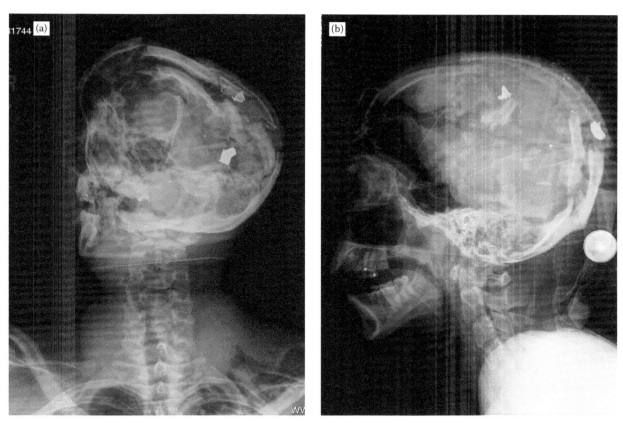

FIGURE 2.3   Anterior-posterior (AP) and lateral views of the skull in a gunshot wound victim; these two views are the standard radiographs obtained for skull injury.

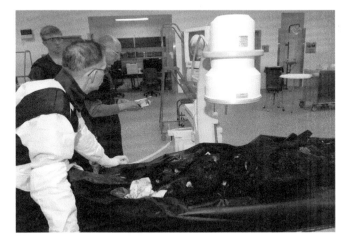

FIGURE 2.4   C-arm fluoroscopy in the autopsy room. The radiologist charts the location of metallic fragments in a severely burned and charred victim of an explosion.

dissection. It may also be used for limited angiographic assessment of vascular integrity by directly injecting the vessel of concern with iodinated contrast material under fluoroscopic observation. The use of C-arm fluoroscopy in a forensic facility should be minimized because it is usually done in a location without physical radiation protection. All personnel in the vicinity of an operating C-arm unit should observe personal radiation safety precautions (e.g., lead aprons, gloves, and thyroid shields) (Figure 2.4). C-arm use dramatically decreased in our facility when MDCT became available. Now there are only a few instances when the C-arm unit is necessary in routine cases. On the rare occasion that both radiography and MDCT are nonoperational, C-arm fluoroscopy may be the only modality available for metal localization. In these cases, serial head-to-toe images are acquired, and the metallic fragments are simultaneously charted. Selected orthogonal views are obtained by rotation of the C-arm when necessary.

Multidetector Computed Tomography

MDCT, also known as multislice computed tomography (MSCT), is used with every autopsy at our institution. It is used as an anatomic survey to evaluate known injury prior to autopsy, to screen nontraumatic deaths for occult injury, and to precisely localize bullets and other metal or foreign objects in three dimensions. It also enables our medical examiners to visualize fractures and fracture patterns in three dimensions, which allows them to study and interpret injury mechanisms with ease.

MDCT scanners have a faster tube rotation speed compared to conventional axial and helical CT scanners. MDCT data are acquired as a volume of data in a single scan by using a two-dimensional array of detector elements rather than the linear array of detector elements used in conventional axial and helical CT scanners. The alignment of the detectors along the length (z-axis) of the body enables the scanners to obtain 4, 8, 16, 64, or more slices with each rotation of the x-ray tube. The data set is viewed as a volume rather than as individual images. The data can be reformatted into two- and three-dimensional planes that closely replicate conventional autopsy. PACS networks are necessary to store and retrieve the data, and postprocessing workstations are necessary to view and manipulate the images for interpretation.

A variety of protocols can be used to obtain scans. Protocols specify the technical and anatomic parameters for obtaining scan data, reconstructing scan data into images, and reformatting images into anatomic planes. Scanning protocols can be organized for specific anatomic regions of the body similar to clinical scanning protocols or can be more generalized to obtain full-body data. Protocols should be established according to the technical capabilities of the specific scanner used.

We scan all of our decedents on a GE Lightspeed 16 Xtra (General Electric [GE] Medical Systems, Milwaukee, Wisconsin). They are scanned with their arms at their sides. Two scans for each body are obtained: a dedicated head CT and a full-body CT (Figure 2.5). The head scans are acquired axially with a slice thickness of 2.5 mm and a slice interval of 2.5 mm, with the CT gantry angled parallel to the orbital-meatal line. The head CT scan data are reconstructed into two separate series, one using a soft tissue algorithm and one using a bone algorithm, and are viewed in multiple window and level settings (Figure 2.6). The total-body scans are obtained from the skull vertex to distal point allowable by table travel (up to 2000 mm). The scanning parameters for the total body scans are as follows: detector configuration 16×0.625, pitch 1.375:1, table speed 13.75 seconds, reconstruction thickness 0.625 mm, and reconstruction interval 0.625 mm. No contrast material is administered. The full-body scans are reconstructed in soft tissue algorithm. Images are viewed on the PACS workstations in multiple window settings to optimize the depiction of pathology and injury (Figure 2.7). Three-dimensional workstations are preferred for image manipulation and reformations. We currently use a GE Advantage Workstation (software version 4.2-07, GE Medical Systems) (Figure 2.2c) and a Vitrea Workstation (software version 4.0.0.0, Vital Images, Inc., Minnetonka, Minnesota). The software developed for image processing in clinical radiology is also suited for postmortem imaging; no special adaptations are required.

Our routine protocol calls for all bodies to be scanned fully clothed with arms down by the sides, in the body bags, as they arrived in the mortuary. This protocol minimizes disturbance of the body and movement of forensic evidence. However, obtaining scans with arms down by the sides creates artifact in the chest and abdominal portions of the scan (Figure 2.8). If the arms are raised, similar to the positioning of clinical body scan, the artifact is removed, and the image resolution improves (Figure 2.8). In some cases, rigor mortis must be broken to raise the arms above the head. If obtaining a scan of the body with the arms above the head is desired and rigor must be broken, consultation with the forensic pathologist should occur.

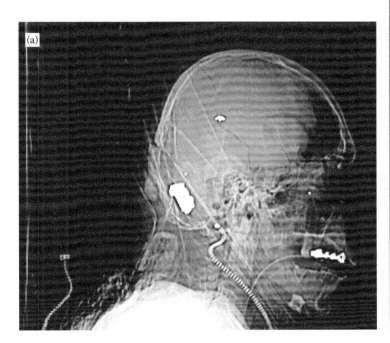

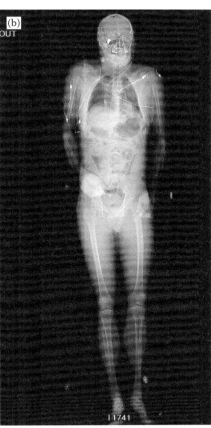

FIGURE 2.5   Routine MDCT protocol prior to autopsy includes dedicated head and full-body CT. (a) Lateral scout view of the head is used to plan the head CT parameters. (b) Anterior-posterior scout of the full body is used to plan the full-body CT parameters.

Images are always viewed in multiple window and level settings to take full advantage of the data provided (Figures 2.6 and 2.7). Acquisition of the total-body scan in 0.0625 mm isotropic voxels enables multiplanar reformations of images that have the same spatial resolution as the original sections without degradation of image quality. Multiplanar and volumetric reformations are quickly and easily generated on state-of-the-art workstations. They enable injury patterns to be visualized in a manner similar to dissection (Figure 2.9 through Figure 2.11). This is particularly helpful to explain and show injury patterns to those who are not facile with axial anatomy. However, some of the two- and three-dimensional software algorithms employed may obscure or mask important anatomic details (Figure 2.10 and Figure 2.12). Therefore, axial images continue to be important reference images that should always be interpreted and referred to because three-dimensional images, maximum intensity projection

(MIP) images, and minimum intensity projection (MinIP) images may not reveal the full extent of injury.

MIP images record the highest attenuation value for a pixel in the path of an x-ray cast through a specified slab of images. MIP images are helpful for highlighting high attenuation structures like bone, bone fragments, or metallic fragments (Figure 2.12). In contrast, MinIP images record the lowest attenuation value for a pixel in a specified slab of images. MinIP images are most valuable for the evaluation of low attenuation structures like air. Therefore, they are very useful to evaluate the trachea and bronchi and the degree of emphysema in lungs (Figures 2.13 and 2.14).

Three-dimensional volume-rendered images are an excellent method to depict specific anatomic abnormalities and fracture patterns. Because routine postmortem CT is performed without intravascular contrast, three-dimensional images are most useful for bone findings such as fractures.

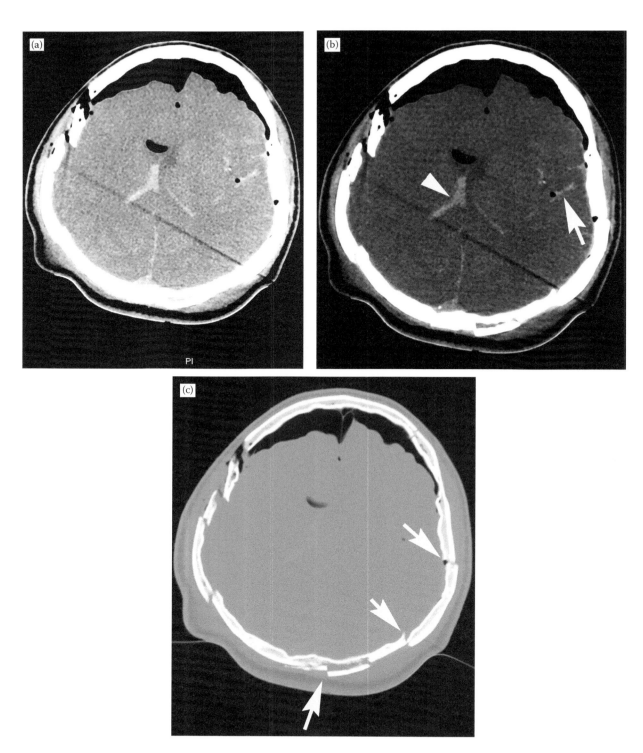

FIGURE 2.6 Head CT scans of a gunshot wound victim are reconstructed into two separate series and viewed in multiple window and level settings. (a) Axial head CT reconstructed in a soft tissue algorithm and viewed with a window and level setting to optimize visualization of the brain parenchyma. (b) The same scan shown in (a) viewed with a window and level setting to optimize the detection of hemorrhage (arrow). Blood is also present in the lateral ventricles (arrowhead on right lateral ventricle). (c) Axial head CT from the same scan data as (a) and (b) is reconstructed in bone algorithm. The bone margin and detail are sharper to more easily identify abnormalities (arrows).

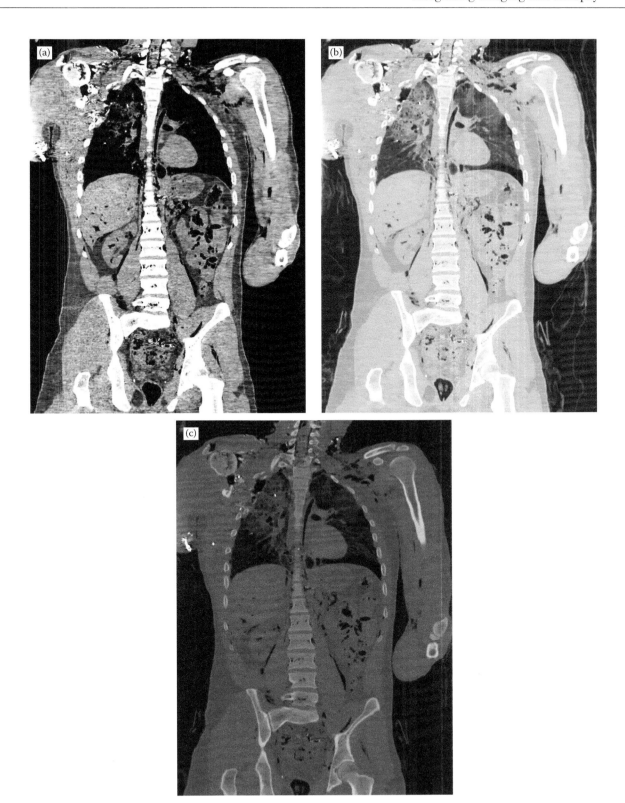

FIGURE 2.7   Coronal CT images of the chest and abdomen in a gunshot wound victim are shown in soft tissue (a), lung (b), and bone (c) window settings. Altering window and level settings allows improved visualization of anatomic structures based upon CT attenuation.

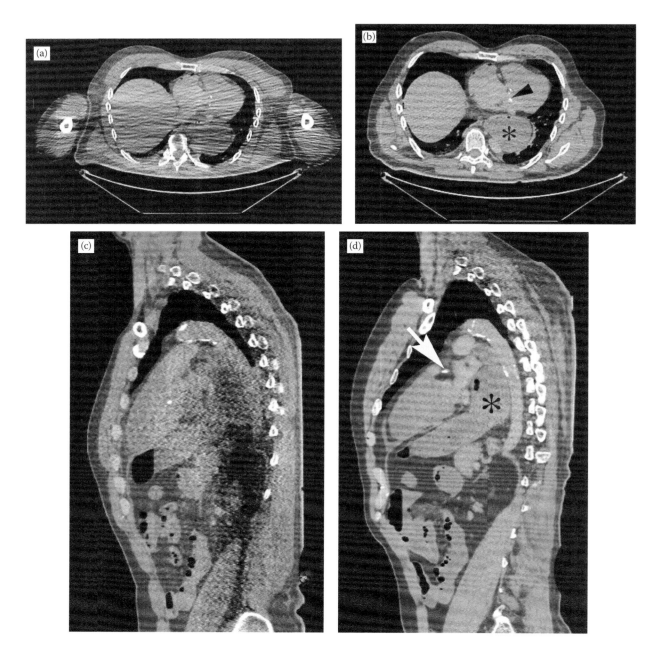

FIGURE 2.8 MDCT images of a man who died from atherosclerotic cardiovascular artery disease. Images were obtained with arms down by the side and arms raised above the head. (a, b) Axial images of the chest show degradation of the image from beam hardening artifact from the arms down in image (a). Artifact is removed when the arms are up in (b). The image shows cardiac calcification (arrowhead), an enlarged right atrium, and a large hiatal hernia (asterisk). (c, d) Coronal MDCT reconstructions from the same scan data show artifact in (c) when the arms are down and improved image resolution in (d) when the arms are up, aorta and coronary artery calcification (arrow) is clearly seen. A hiatal hernia is apparent behind the heart and there are aortic calcifications.

We have found that the three-dimensional depictions of skull, spine, and pelvic fractures are very useful because it is often difficult to visualize these fractures at autopsy (Figure 2.15). Skin surface–rendered images are not as useful because current software does not generate images that fully characterize the margins and contours of abnormalities or defects in the skin. Furthermore, contusion, hemorrhage, and other findings like soot deposition or superficial marking of the skin cannot be evaluated on these images.

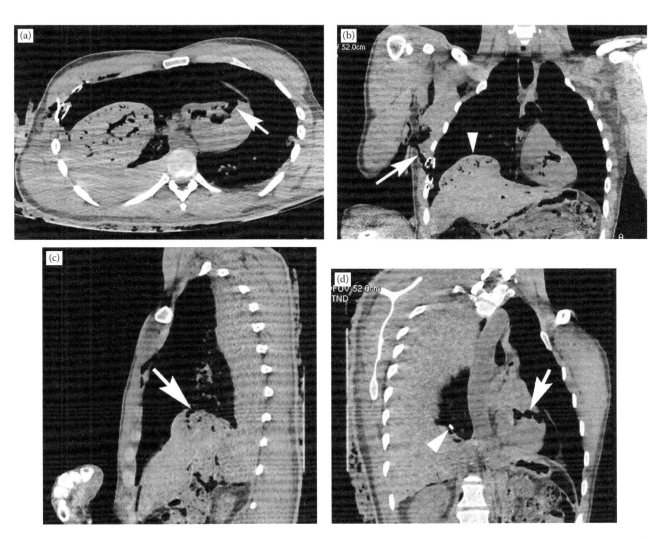

FIGURE 2.9   Multiplanar MDCT images of the chest in a gunshot wound victim allow display and evaluation of anatomic findings in multiple planes. (a) Axial image shows a right rib fracture and right hemothorax. The bullet passed through the heart (arrow). There is pneumopericardium and elevation of the right hemidiaphragm and liver into the right hemithorax. (b) Coronal image shows the gunshot wound track in the right chest wall (arrow), rib fracture, and right diaphragmatic hernia (arrowhead). (c) Sagittal image shows the bullet passed through the right hemidiaphragm (arrow). Hemothorax layers dependently in the posterior right hemithorax. (d) Coronal oblique image shows the gunshot wound track through the heart (arrow). A bullet fragment is present in the right lung (arrowhead). Pneumopericardium and hemothorax are again identified.

Angiography and Multidetector Computed Tomography Angiography

One of the most important limitations of routine postmortem imaging is the assessment of vascular injury and pathology. A variety of postmortem angiography techniques have been reported (Grabherr et al. 2006, Grabherr et al. 2007, Ross et al. 2008a, 2008b). The contrast agent, delivery mechanism, and injection technique can be altered based upon the location of the suspected abnormality and radiologic technique

used to image during contrast injection. For example, the injection may be performed during autopsy with C-arm fluoroscopy; before autopsy with radiography or MDCT; or after autopsy cannulation of vessels with radiography, C-arm fluoroscopy, or MDCT (Harcke and Solomon 2008).

Postmortem angiography differs from antemortem techniques in that arterial perfusion can be either antegrade or retrograde to physiologic flow. This has enabled us to study a region of interest by gaining access at the most

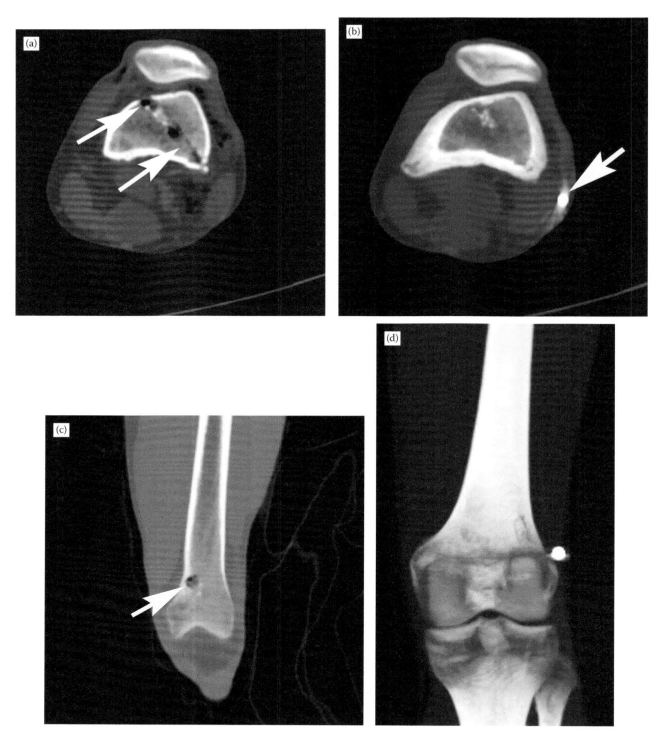

FIGURE 2.10   Multiplanar and three-dimensional imaging of a projectile injury through the distal femur. Axial MDCT (a) and maximum intensity projection (b) images of the distal femur show a wound track through the distal femur (arrows in a) and a metallic bullet fragment in the soft tissues lateral to the femoral condyle (arrow in b). Decompositional gas is present in the soft tissues and vascular structures. (c) Coronal MDCT of the distal femur shows the entry site of the fragment in the medial aspect of the distal femur (arrow). (d) Maximum intensity projection (MIP) image of the distal femur and knee shows the location of the metallic fragment with respect to the bone and joint. Importantly, the wound track is not clearly identified on the MIP images, emphasizing the importance of the routine images.

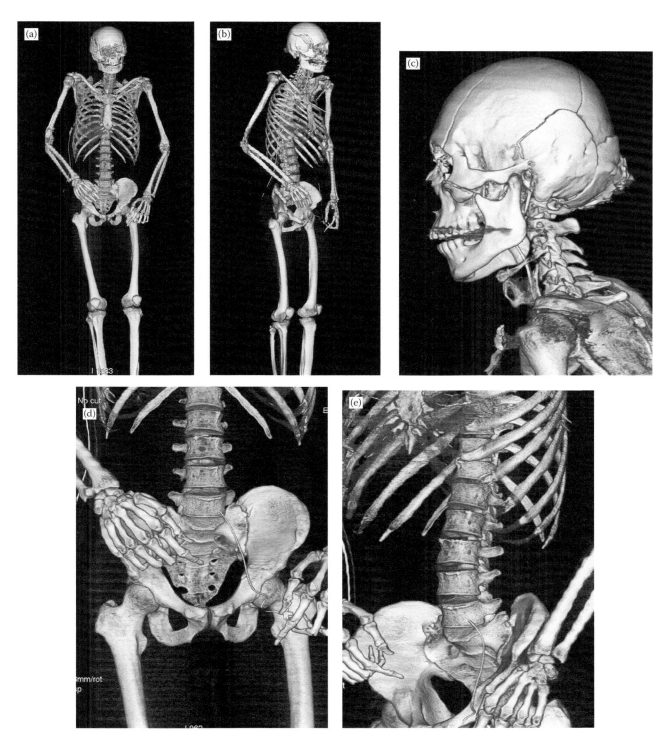

FIGURE 2.11   Three-dimensional MDCT of a victim of head trauma shows frontal (a) and oblique (b) total-body images that are optimized to view bones. Images may be manipulated to magnify abnormalities without resolution loss. The skull (c), spine, and pelvis (d, e) may be rotated to view all sides. Skull fracture detail is well seen in (c), and a left femoral venous catheter is seen in (d) and (e). A right-sided chest tube is present.

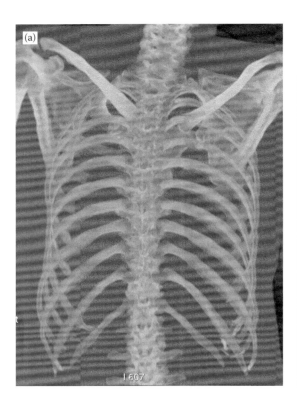

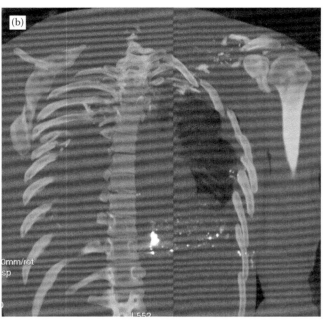

FIGURE 2.12   Maximum intensity projection (MIP) images of the thorax. (a) MIP image of the bony thorax from a man who died of atherosclerotic coronary vascular disease shows no bone abnormalities. (b) MIP image of the bony thorax in a gunshot wound victim shows multiple fractures, bone fragments, and metallic fragments in the gunshot wound tracks. Unlike the image shown in (a), the entire slab of images was not used to create this MIP. Portions of the lung and heart are present in the image providing an unusual view of the thorax. Care must be taken when creating these images such that the true anatomic findings are not misrepresented.

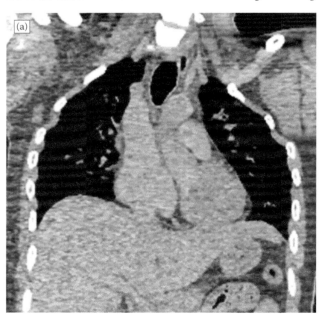

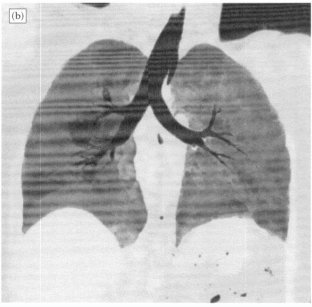

FIGURE 2.13   Minimum intensity projection (MinIP) images of the thorax in a man who died from atherosclerotic cardiovascular disease. (a) Coronal MDCT of the chest shown in soft tissue windows shows an enlarged right atrium and prominent superior vena cava. (b) Coronal MinIP shown in lung windows shows a normal trachea and central bronchi with an asymmetric distribution of pulmonary edema in the lungs. The lucent regions are better aerated than the more attenuated regions.

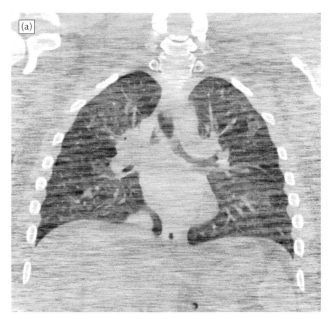
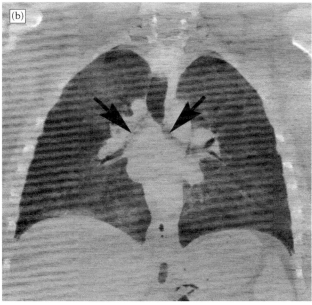

FIGURE 2.14 Minimum intensity projection (MinIP) images of the thorax in a drowning victim. (a) Coronal MDCT of the chest shows pulmonary edema and abnormal density in the trachea and bronchi. The MinIP image in (b) shows the density in the airways represents froth (arrows).

convenient location and time. For example, we perform vertebral arteriograms after removal of the brain by cannulating the intracranial vertebral artery and injecting contrast in retrograde manner (Figure 2.16). Our angiographic technique employs a 50% dilution of a standard iodinated contrast (320 mg of iodine/mL) that is hand injected. The flexibility in when and how postmortem angiography can be performed allows the pathologist to ask for the procedure at a point in the autopsy when it can be done with minimal delay in the procedure. The short time required for an angiogram is in contrast to the time for a difficult and lengthy dissection.

Magnetic Resonance Imaging

MRI has superior contrast resolution compared to MDCT. Consequently, it is a useful technique to image soft tissue alterations and pathologic processes. Postmortem MRI has been used to assess soft tissue and visceral hemorrhage, ischemia, and tumors (Aghayev et al. 2008, Ikeda et al. 2007, Shiotani et al. 2005, Yen et al. 2007). However, the technical complexity and availability of MRI make it more complicated to use as a routine imaging modality compared to MDCT. Procedures for cases with ferromagnetic ballistic fragments also need to be established.

Ultrasound

Although not as commonly used in forensic medicine as MDCT and MRI, ultrasound has been reported to be an effective imaging modality in autopsy diagnosis (Akopov et al. 1976, Uchigasaki et al. 2004). Ultrasound is less expensive and more readily available than MDCT and MRI. It has the potential to be a useful tool when performing limited autopsy or to guide organ biopsy in a minimally invasive autopsy. A major limitation of sonography is the degradation of images by the artifacts produced by air and gas. Soft tissue gas is a major feature of decomposition, and this has the likelihood to limit postmortem use.

RADIOLOGY REPORTING

The determination of the cause and manner of death is decided by the medical examiner and reported in the final autopsy report. The final autopsy report includes data that are obtained from many supporting elements of the entire forensic examination: scene investigation; medical and social history; fingerprint and DNA analysis; forensic dental examination; serum, urine, body fluid biochemical and toxicological analysis; external examination; photographic documentation;

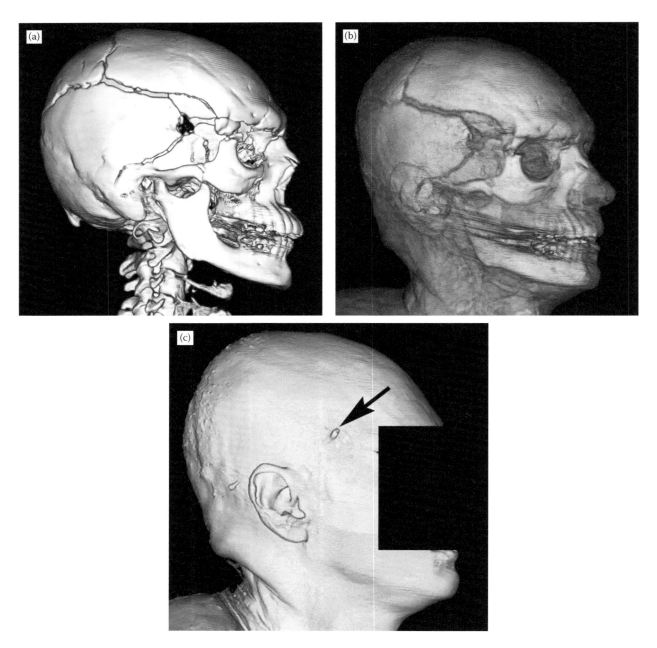

FIGURE 2.15 Three-dimensional images of the head in a gunshot wound victim generated with different workstation protocols show various aspects of the injury. (a) Three-dimensional volume–rendered images of the bones of the skull and neck show a complex skull fracture with depressed fragments and a radial orientation of the fracture lines from the bullet entry site. (b) Three-dimensional soft tissue and bone volumetric image shows soft tissue overlying the bone injury. (c) Three-dimensional surface–rendered image shows the gunshot wound entry site (arrow) in the skin.

internal dissection; histologic tissue evaluation; and, the radiology report. In our practice, the radiologist or forensic pathologist may complete the radiology report depending on the personnel support available at the time of autopsy. In the most optimal practice situation, images are reviewed by the radiologist (either onsite or remotely) and discussed with the forensic pathologist so the most complete imaging information is available before autopsy. If the radiologist is not available at the time of autopsy, the forensic pathologist may choose to preliminarily chart the imaging findings before autopsy and have the radiologist compile a formal report at a

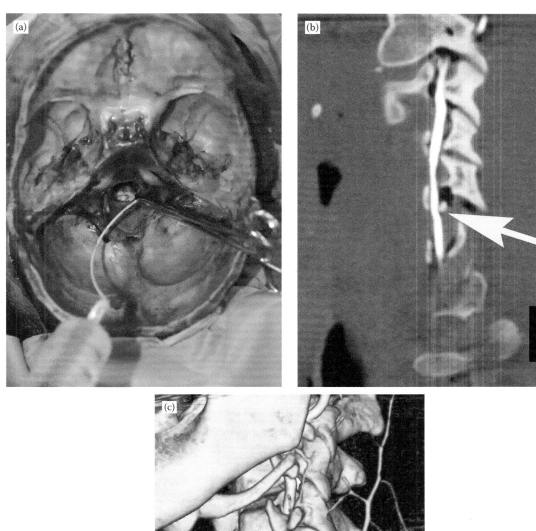

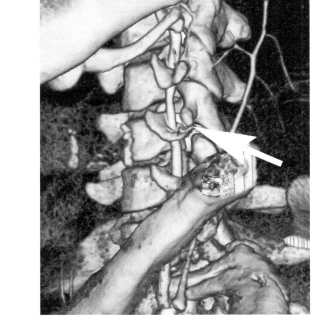

FIGURE 2.16 Vertebral angiography in a blunt trauma victim with a fracture of the left transverse process of C-5. (a) Autopsy photograph shows cannulation of a vertebral artery in preparation for a retrograde injection of contrast. (b) Sagittal MDCT obtained during contrast injection at the plane of the vertebral foramen shows a normal vertebral artery. The fracture fragment presses on the artery (arrow). (c) Three-dimensional MDCT shows the normal contrast-filled vertebral artery passing through the vertebral foramen and the transverse process fracture (arrow).

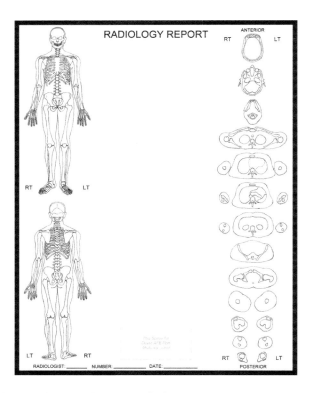

FIGURE 2.17   Anatomic worksheet used to record radiology findings.

later time. Most importantly, the imaging studies should be reviewed prior to dissection so the forensic pathologist is aware of pertinent findings and can tailor the autopsy for such things as recovery of metal fragments, extent of injury, and assessment of prior medical intervention.

As cross-sectional imaging becomes more widespread in forensic medicine and more radiologists become involved in forensic medicine, it becomes clear that the radiologist will encounter similar issues that are present in clinical medicine, such as what is the best way to communicate in a timely manner and how much historical information is necessary when reading the studies. Retaining objectivity in the interpretation is important because the interpretation and report may have to be explained and defended in the courtroom. However, the collaboration between the radiologist and forensic pathologist strengthens the conclusions of the final autopsy findings. We found that balancing objectivity with knowledge of the autopsy can be difficult. Our preference is

to initially view and interpret postmortem images with basic historical data only so we are not biased by preceding investigative reports. We like to discuss the findings with the medical examiner prior to autopsy and then arrive at a radiologic and pathologic consensus after dissection is complete. In our situation, the location of radiology adjacent to the autopsy room facilitates this process and has helped us learn the strengths and limitations of postmortem MDCT.

The format of the radiology report may be similar to a clinical radiology report or may be charted anatomically similar to other forensic worksheets. We adapted the latter in our practice. The radiology findings of a combined interpretation of radiography and MDCT are typically annotated on a drawing of the human body and schematic CT sections so the medical examiner can use the data during autopsy (Figure 2.17). This worksheet joins the other worksheets generated by the pathologist and used in composing the final autopsy report.

CONCLUSIONS

Incorporation of cross-sectional imaging modalities such as MDCT into a forensic facility requires careful consideration of the impact on daily workflow. In our practice, cross-sectional imaging augments standard radiography. If cross-sectional imaging is performed on every case, consistency in the workflow and imaging protocols is essential to avoid error.

REFERENCES

Aghayev, E., Thali, M. J., Jackowski, C. et al. 2008. MRI detects hemorrhages in the muscles of the back in hypothermia. *Forensic Sci Int* 176: 183–186.

Akopov, V. I., Lozovskii, B. V., and Kuryshev, A. N. 1976. Possibilities of using ultrasonic diagnosis in forensic medicine. *Sud Med Ekspert* 19: 16–18.

Brogdon, B. G. 1998. *Forensic radiology,* Boca Raton, FL: CRC Press.

Fatteh, A. V., and Mann, G. T. 1969. The role of radiology in forensic pathology. *Med Sci Law* 9: 27–30.

Grabherr, S., Djonov, V., Friess, A. et al. 2006. Postmortem angiography after vascular perfusion with diesel oil and a lipophilic contrast agent. *AJR Am J Roentgenol* 187: W515–W523.

Grabherr, S., Djonov, V., Yen, K., Thali, M. J., and Dirnhofer, R. 2007. Postmortem angiography: review of former and current methods. *AJR Am J Roentgenol* 188: 832–838.

Harcke, H. T., and Solomon, C. 2008. Postmortem angiography in support of radiology assisted autopsy. American Academy of Forensic Sciences 60th Annual Scientific Meeting, Washington, DC.

Ikeda, G., Yamamoto, R., Suzuki, M. et al. 2007. Postmortem computed tomography and magnetic resonance imaging in a case of terminal-stage small cell lung cancer: an experience of autopsy imaging in tumor-related death. *Radiat Med* 25: 84–87.

Mann, G. T., and Fatteh, A. B. 1968. The role of radiology in the identification of human remains: report of a case. *J Forensic Sci Soc* 8: 67–68.

Ross, S., Spendlove, D., Bolliger, S. et al. 2008a. Postmortem whole-body CT angiography: evaluation of two contrast media solutions. *AJR Am J Roentgenol* 190: 1380–1389.

Ross, S., Spendlove, D., Bolliger, S., Oesterhelweg, L., and Thali, M. 2008b. Postmortem minimal invasive CT-angiography: the next step toward a virtual autopsy. Radiologic Society of North America 94th Scientific Assembly and Annual Meeting, Chicago, IL.

Shiotani, S., Yamazaki, K., Kikuchi, K. et al. 2005. Postmortem magnetic resonance imaging (PMMRI) demonstration of reversible injury phase myocardium in a case of sudden death from acute coronary plaque change. *Radiat Med* 23: 563–565.

Uchigasaki, S., Oesterhelweg, L., Gehl, A. et al. 2004. Application of compact ultrasound imaging device to postmortem diagnosis. *Forensic Sci Int* 140: 33–41.

Yen, K., Vock, P., Christe, A. et al. 2007. Clinical forensic radiology in strangulation victims: forensic expertise based on magnetic resonance imaging (MRI) findings. *Int J Legal Med* 121: 115–123.

Chapter 3

Postmortem Change and Decomposition

FORENSIC PRINCIPLES

Postmortem change and decomposition begin to develop immediately upon death. Postmortem change most commonly refers to livor mortis (also known as lividity or postmortem hypostasis), rigor mortis, and algor mortis. The degree of postmortem change and decomposition in combination with observations at the scene of death and, in rare cases, the status of gastric digestion can be used to estimate the time of death. They may also be indicators of the position of the body at the time of death or whether a body has been moved or tampered with after death.

Livor mortis is the earliest visible postmortem change. When the circulatory system stops, active blood flow through capillary beds ceases, and gravity acts upon the stagnant blood to draw it to the most dependent locations of the body. Erythrocytes are the most affected components of blood and account for the appearance of the bluish red or reddish purple coloration of livor mortis (Knight and Saukko 2004). The distribution and location of livor mortis depend on the position of the body after death. For example, when a body is supine following death, livor mortis is seen in the posterior regions of the body. In a body hanging vertically, livor mortis will be most marked in the feet and lower legs.

Decomposition is a multifactorial process that involves cellular autolysis, bacterial fermentation, and insect and animal predation. The rate and extent of postmortem change and decomposition are quite variable and dependent upon the cause of death and external environment. Most importantly, decomposition may dramatically alter the appearance of the body such that it may make determining the cause of death more difficult.

AUTOPSY FINDINGS

Livor Mortis

The gradual appearance, concentration, and fixation of livor mortis are due to progressive intravascular hemo-concentration and lysis of the erythrocytes. Initially, livor mortis appears as small reddish purple cutaneous patches, which gradually coalesce. The distribution of livor mortis will shift with repositioning of the body and will blanch with applied pressure. As the postmortem period increases in time, livor mortis will become less movable and, eventually, will become fixed and unmovable. Although the onset and fixation of livor mortis may be affected by body size, cause of death, position of the body, and ambient temperature, it is generally accepted that livor mortis becomes visible from 20 minutes to 2 hours after death and reaches its maximum at 8 to 12 hours after death, at which time it becomes fixed (Di Maio and Di Maio 2001, Payne-James et al. 2003). The dependent areas supporting the weight of the body, such as the upper back, buttocks, and calves in a supinely positioned body, will appear paler compared to the surrounding livor mortis because body weight compresses blood vessels in these areas and prevents the accumulation of erythrocytes (Figures 3.1 and 3.2). Blanched areas may also appear in regions that are compressed from tight clothing, such as the straps from a brassiere or tight elastic in socks.

The color of livor mortis may vary with the degree of hemoglobin oxygenation at the time of death. The usual color is bluish or purplish red. In general, those who die hypoxic have darker livor mortis. For example, many natural deaths from atherosclerotic coronary artery disease or congestive heart failure have a dark livor mortis that is more prominent in the upper body and face. The color

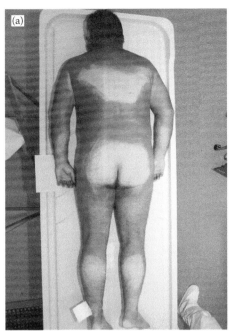

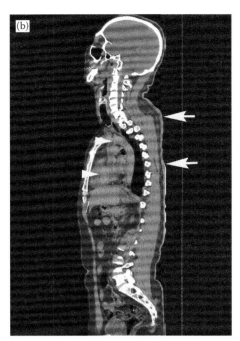

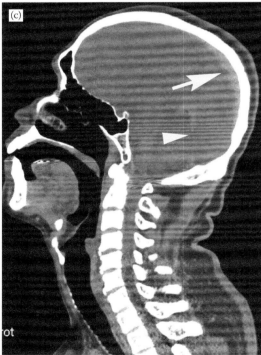

FIGURE 3.1   Postmortem livor mortis in a man who died from atherosclerotic coronary artery disease and was found in the supine position. (a) Autopsy photograph shows red and bluish red livor mortis on the posterior region of the body. Focal areas on the upper back, buttocks, calves, and heels are pale and blanched because these portions of the body supported the weight of the body. (b) Sagittal MDCT at the level of the aortic arch shows thickening and increased attenuation of the skin and subcutaneous fat (arrows) from livor mortis in the posterior region of the body and a hematocrit effect of fluid levels in the aorta and cardiac chambers (arrowheads). (c) Sagittal MDCT image of the brain shows high attenuation hemoconcentration in the superior sagittal sinus (arrow) and transverse sinus (arrowhead). Note the presence of skin thickening posteriorly.

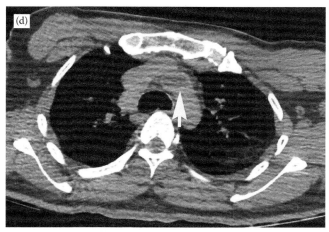

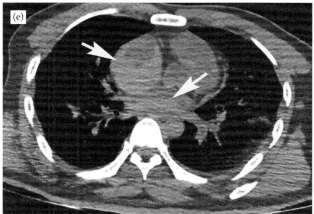 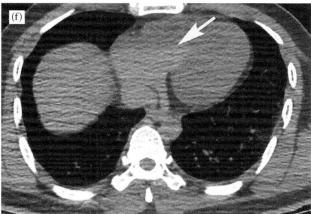

FIGURE 3.1 (*Continued*) Postmortem livor mortis in a man who died from atherosclerotic coronary artery disease and was found in the supine position. (d, e, f) Axial MDCT images show high attenuation hemoconcentration (arrows) in the dependent portions of the aorta, right and left atria, and right ventricle, which form a hematocrit effect. The anterior aortic wall appears relatively dense adjacent to the hypoattenuating serum component of blood.

of livor mortis may also be indicative of specific etiologies of poisoning: cherry red in carbon monoxide poisoning, hypothermia, or cyanide poisoning; brown from the methemoglobin formed in sodium nitrite poisoning; and green from sulfhemoglobin in hydrogen sulfide poisoning (Payne-James et al. 2003).

Livor mortis is a phenomenon that occurs throughout all fluid compartments, tissues, and organs in the body. Internally, the lungs often show the most marked changes on gross inspection. The nondependent portions of the lungs, most commonly anterior, are pale, and the dependent posterior lung is dark bluish red (Figure 3.3). Similar changes may be seen in other organs, such as the dependent, pelvic segments of jejunum and ileum, which may show areas of patchy discoloration from livor mortis. Similar to

the cutaneous manifestations of livor mortis, visceral livor mortis is secondary to concentration of erythrocytes within the vascular system and diffusion of plasma from the vascular system (Ambrosi and Carriero 1965).

Rigor Mortis

Immediately after death, the muscles of the body are flaccid. Within 1 to 6 hours, skeletal muscles become increasingly stiff and rigid due to rigor mortis, by 6 to 12 hours after death, muscles are completely rigid, and rigor mortis is fully developed. Biochemically, rigor mortis is the formation of a transient bond between actin and myosin filaments that occur because of the loss of adenosine triphosphate (ATP) from muscle. Muscle groups will remain in full rigor until they are forcibly moved (*breaking rigor*), which physically ruptures muscle fibers

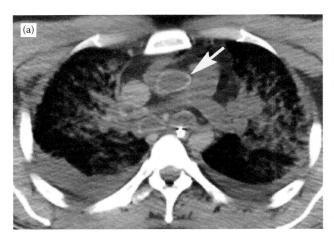

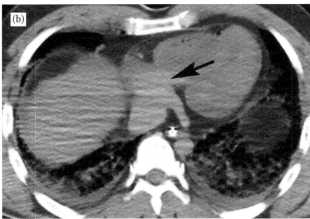

FIGURE 3.2 Postmortem livor mortis in a man who died from atherosclerotic coronary vascular disease. (a, b) Axial MDCT images of the chest show the superior vena cava and descending aorta have higher attenuation compared to the ascending aorta (arrow) in (a). The wall of the ascending aorta appears denser than the blood in the lumen because the red blood cells have settled to the more dependent vessels. High attenuation–dependent hemoconcentration is present in the right atrium and inferior vena cava (arrow in b). There is pulmonary edema, bilateral pleural effusions, and a nasogastric tube in the esophagus.

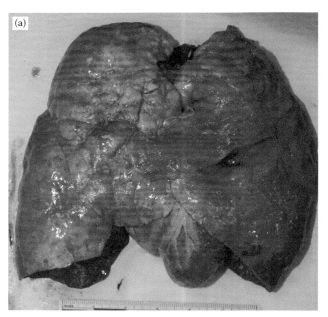

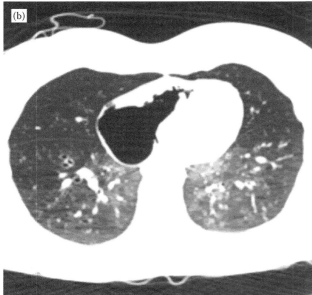

FIGURE 3.3 Pulmonary livor mortis. (a) Autopsy photograph shows pale red anterior lungs and a dark red posterior lung from livor mortis. (b) Axial MDCT shows a gradual increase in pulmonary attenuation from anterior to posterior. Decompositional gas is present in the right heart.

and tears tendon insertions, or until autolysis and early decomposition causes the muscles to loosen, which usually occurs approximately 24 to 36 hours at ambient temperature. It is difficult to rely solely on the stage of rigor mortis to estimate the time of death because there are many factors that affect the onset of rigor mortis: internal body temperature, decedent's activity prior to death, environmental temperature, and the amount of muscle mass within an individual (Dix and Graham 2000). In some cases, rigor mortis may be helpful to determine the position of the body at the time of death or if a body has been moved after death.

Algor Mortis

Algor mortis, or body cooling, is the equilibration of the body temperature with the surrounding environment after death. Because there are a number of variables that affect body cooling, algor mortis alone is not a useful predictor of the time of death. However, body temperature may be useful in combination with other postmortem findings to estimate the time of death.

Decomposition

Decomposition is the breakdown of dead soft tissues into fluids and gases. It is a multifactorial process that occurs through autolysis, putrefaction, and in some cases, insect activity and animal predation. Postmortem autolysis is the breakdown of individual cells by intracellular proteolytic lysosomal enzymes. The rate of postmortem autolysis is greatly affected by temperature. Warmth and humidity accelerate autolysis, cold temperatures slow autolysis, and autolysis is stopped by freezing (Di Maio and Di Maio 2001). Putrefaction is a complex chemical process caused by bacteria and fermentation. The normal bacterial flora of the gastrointestinal tract as well as bacteria that are pathologically present within the body spread throughout the body to produce putrefaction.

The rate of decomposition is affected by a number of interrelated variables that include ambient temperature, humidity, body size and weight, trauma to the body, clothing, the surface or soil the body is resting on, and access of the body to insects and other predators (Mann et al. 1990). Consequently, the rate of decomposition is inherently unpredictable. There are no scientifically based methods to estimate the time of death of human bodies based upon the degree of decomposition for a body found in natural settings at ambient temperature. Despite the unpredictable rate and progression of decomposition, putrefactive changes generally follow a predictable sequence. The first external sign of putrefaction in a body that has not sustained external injury is usually patchy green discoloration on the lower abdominal wall (Figure 3.4). Generally, this begins in the right lower

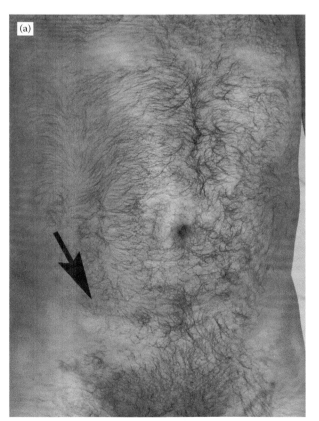 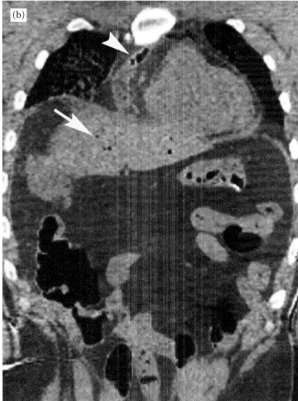

FIGURE 3.4 Early decompositional changes in the abdomen 2 days postmortem. (a) Autopsy photograph of the anterior abdominal wall shows greenish brown discoloration (arrow) from early putrefaction. (b) Coronal MDCT shows putrefactive gas in the portal veins (arrow) and systemic vasculature (arrowhead).

quadrant because of the relatively superficial location of the cecum, which contains large quantities of bacteria. The patchy green discoloration of the lower abdominal wall progresses to a dark green discoloration of the upper torso, head, and neck. The next phase is swelling of the face and head due to gas formation by putrefactive bacteria. Marbling begins to appear at this time (Figure 3.5). Marbling is intravascular hemolysis with a reaction between hemoglobin and hydrogen sulfide that results in a greenish black discoloration of the vessel walls and adjacent tissues. It produces a pattern of branching and arborizing greenish black lines that are visible through the skin. Generalized bloating is the next phase, followed by blister and vesicle formation, loosening of the epidermis, and skin and hair slippage (Figures 3.5 and 3.6). Bloating is frequently more severe in the face and scrotum.

Internally, when the body is intact, decomposition is not as dramatic and rapid as it is on the external surface of the body. Internal organs decompose at different rates. Autolysis of the pancreas and adrenal glands begins

almost immediately after death (Figure 3.7 through Figure 3.9). Likewise, the brain is very vulnerable to autolysis and becomes soft to the touch within days after death and may liquefy within 1 week of death (Figure 3.10 through Figure 3.13) (Knight and Saukko 2004). In contrast, the heart, prostate, and uterus are relatively resistant to autolysis and are usually the last remaining recognizable organs.

Decomposition is not always uniform. Asymmetric or focal acceleration or deceleration of decomposition may occur when a body is clothed, positioned, or sheltered in a manner that creates a microenvironment that favors or slows decomposition (Figure 3.14) (Dix and Graham 2000, Mann et al. 1990). Decomposition also occurs more rapidly at sites of injury where bacteria have been introduced into the body by the injury or insects may gain access (Figure 3.15).

Insect activity, including that of maggots, beetles, and ants, as well as animal predation accelerate decomposition. The location of a body is the major determinant

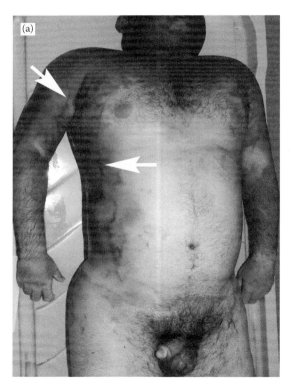 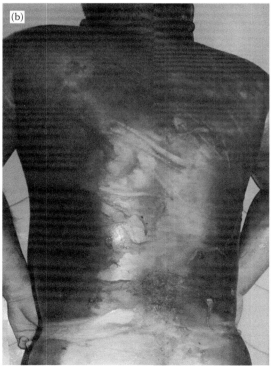

FIGURE 3.5 Moderate decomposition in a drowning victim found 3 days after a witnessed drowning death. (a, b) Autopsy photographs show dark green discoloration and marbling (arrows) of the torso. Bloating and gaseous distension of the head and scrotum are present.

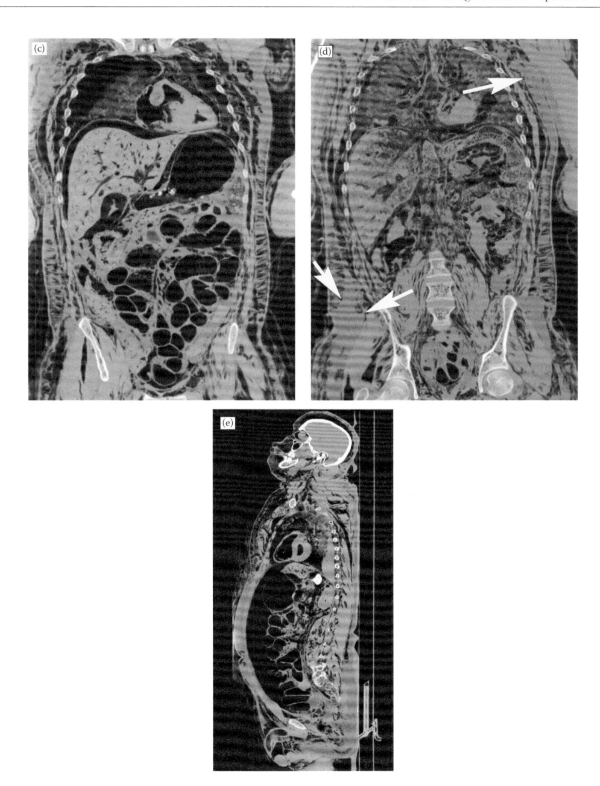

FIGURE 3.5 (*Continued*) Moderate decomposition in a drowning victim found 3 days after a witnessed drowning death. (c, d) Coronal MDCT shows putrefactive gas throughout all vasculature, tissue planes, and rounded foci of gas in the visceral organs. Low attenuation liquefaction is present in the subcutaneous fat (arrows). (e) Sagittal MDCT shows a large amount of gas in the peritoneal cavity, bowel distension, and gaseous distension of the soft tissue of the face, anterior body wall, and scrotum. Ingested sand is in the stomach. Pleural effusions, intraperitoneal fluid, and subcutaneous liquefaction are also present.

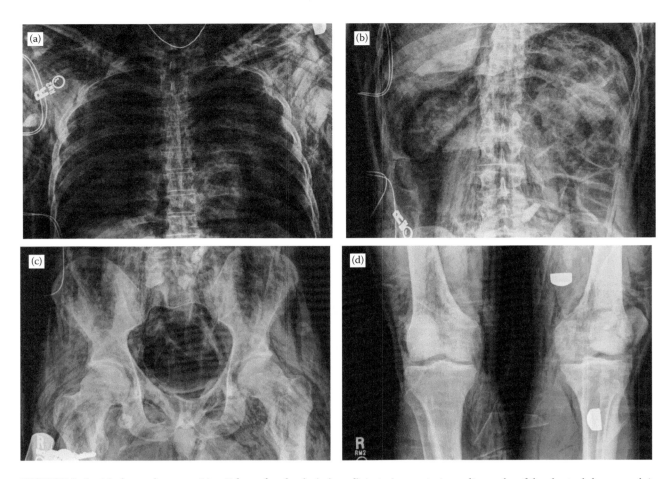

FIGURE 3.6   Moderate decomposition 5 days after death. (a, b, c, d) Anterior-posterior radiographs of the chest, abdomen, pelvis, and lower extremities show symmetrically distributed gas throughout the soft tissue planes and body cavities.

of exposure to insects and carnivores. In criminal investigations, recognition and evaluation of the type of insect and carnivorous activity may aid in a death investigation. Insect and animal predation should not be mistaken for antemortem injury (Figure 3.16 through Figure 3.18). Ligaments, cartilage, and skeletal tags are usually the last remaining soft tissue structures, and they may remain intact for a long time. Eventually, complete skeletalization will occur in the majority of bodies. The time to skeletalization is quite variable and dependent on the factors outlined above. In general, in a hot, dry climate, a body will mummify rather than skeletonize, whereas complete skeletalization may occur in as little as a week in very hot, humid climates (Di Maio and Di Maio 2001).

In specific environmental conditions, bodies will undergo transformation to adipocere or mummify rather than decompose and skeletonize. Adipocere is an important postmortem change that results from the hydrolysis and hydrogenation of adipose tissue. Anaerobic bacteria such as *Clostridium perfringens* facilitate the chemical process that results in the formation of a waxy substance composed of palmitic, oleic, and stearic fatty acids (Knight and Saukko 2004). Adipocere formation tends to occur in bodies that are in moist graves, vaults, or crypts. In general, adipocere formation mainly occurs in the subcutaneous fat, but it may also occur in intra-abdominal mesenteric and retroperitoneal fat. Adipocere formation may coexist with putrefactive decomposition (Figure 3.19). Mummification occurs in dry environments where bacterial growth is inhibited. It may occur in dry or cold environments because the most important factor is the presence of currents of dry air. From a forensic standpoint, mummification and adipocere formation may preserve injuries such as bullet holes on the surface of the body.

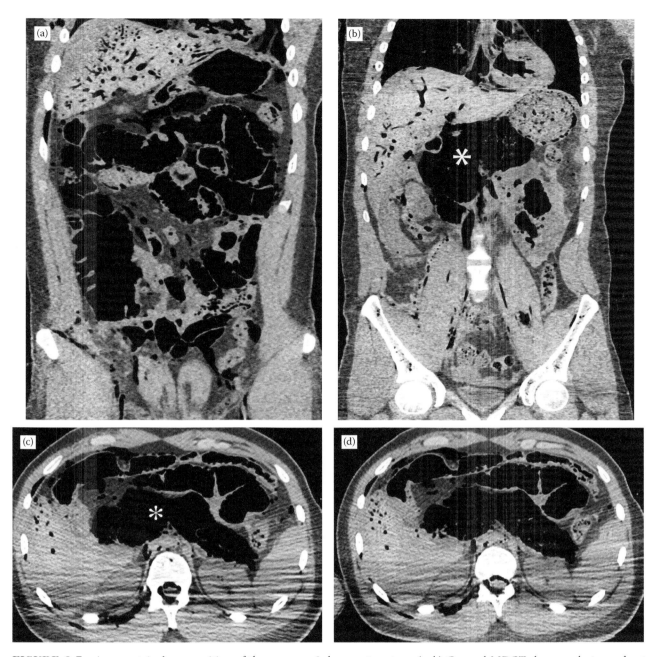

FIGURE 3.7 Asymmetric decomposition of the pancreas 3 days postmortem. (a, b) Coronal MDCT shows early to moderate decomposition in the abdomen and pelvis. There is gaseous distension of the bowel and intravascular gas. The pancreas (asterisk) contains more decompositional gas compared to adjacent visceral organs. (c, d) Axial MDCT shows similar visceral and pancreatic (asterisk) findings as well as fluid around the liver.

Decomposition poses many challenges for forensic pathologists, because findings that may have been diagnostic of the cause of death at the time of death may be more difficult to recognize or disappear completely with decomposition. Injury to soft tissue may not be apparent at dissection because of soft tissue collapse and liquefaction. Presumptive identification is also more challenging, particularly in skeletalized remains. Skeletonized remains are also often scattered by animal predation, making full recovery difficult, if not impossible, in some cases.

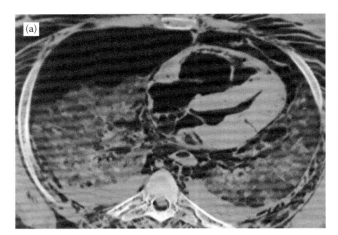

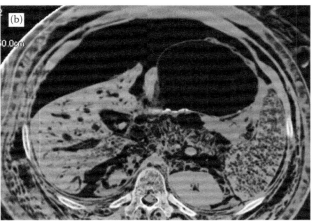

FIGURE 3.8   Moderate to advanced decompositional changes in the chest and abdomen 5 days postmortem. (a, b) Axial MDCT shows partial collapse of the lungs and small bilateral pleural fluid collections secondary to decomposition. Gas is present in the pleural, pericardial, peritoneal, and retroperitoneal spaces, as well as within the heart, blood vessels, connective tissues of the mediastinum, and subcutaneous tissue. Round lucencies of decompositional gas in the spleen and liver and linear collections of gas in the pancreas are typical features of moderate to advanced decomposition.

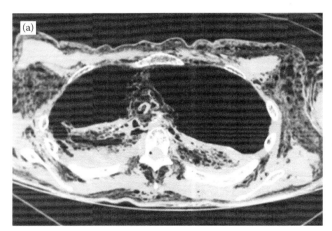

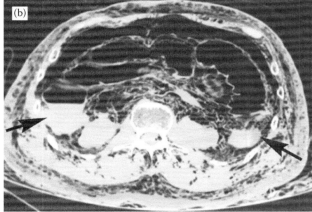

FIGURE 3.9   Advanced decomposition of the chest and abdomen 12 days postmortem. (a) Axial MDCT shows gas throughout all tissue planes. The lungs have collapsed, and there are small effusions and a gas-filled pleural cavity. (b) Intraperitoneal gas is present in the abdomen along with marked intestinal distension and gas in all peritoneal and retroperitoneal tissues and organs. There is a small amount of fluid in the dependent recesses of the peritoneum (arrows in b).

RADIOLOGIC PRINCIPLES

Postmortem change and decomposition are always present at autopsy and on postmortem MDCT because they begin to occur immediately upon death. Consequently, the appearance of postmortem change and decomposition on postmortem MDCT can be considered normal and should not be mistaken for a pathologic process or injury. Postmortem change and decomposition are important findings on MDCT because they may obscure soft tissue injury or pathology, thereby limiting the MDCT assessment of soft tissue for hemorrhage, laceration, and wound tracks in cases with suspected trauma. Putrefactive gas should not be mistaken for pathologic gas collections that may have contributed to death or causes of death such as air embolism, pneumothorax, pneumoperitoneum, or gas-forming infections. Gas in anatomic spaces and in blood vessels can generally be considered putrefactive when it is present symmetrically throughout

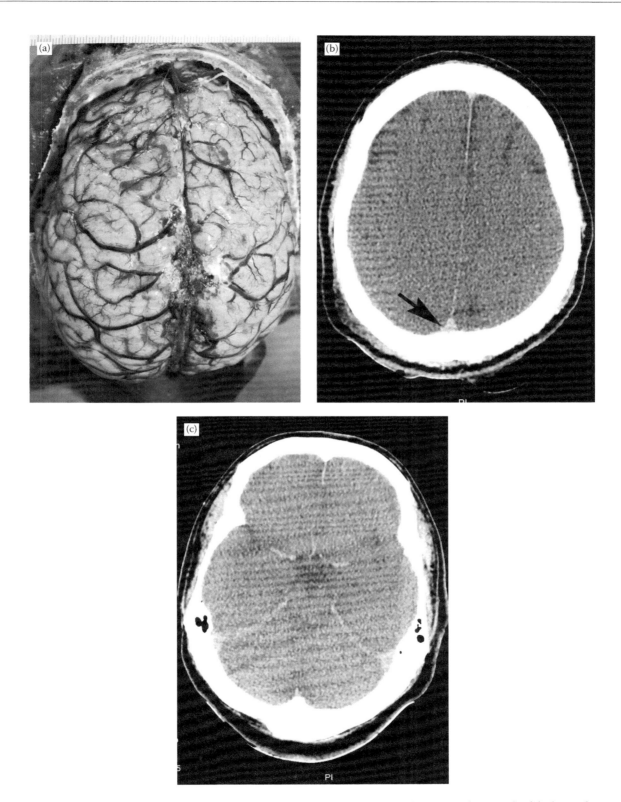

FIGURE 3.10   Early decompositional changes in the brain 48 hours postmortem. (a) Autopsy photograph of the brain shows preservation of gyri and coloration. (b) Axial MDCT shows effaced sulci and ventricles and complete loss of gray-white differentiation in the cerebral hemispheres. There is increased attenuation from livor mortis in the posterior sagittal sinus (arrow). (c) Axial MDCT at the level of the mesencephalon shows increased attenuation in middle and anterior cerebral arteries. Note that the tentorium and anterior portion of the falx are also high attenuation areas.

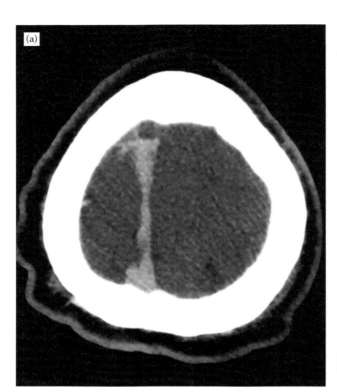

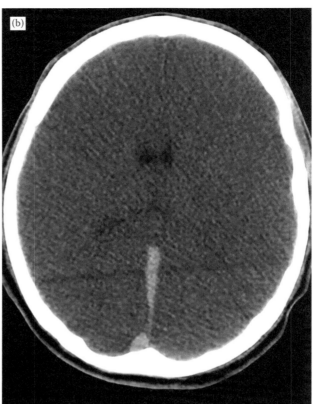

FIGURE 3.11   Early decomposition and livor mortis in the brain 2 days postmortem. (a, b) Axial MDCT of the brain shows loss of differentiation between the gray and white matter and compression of the ventricular system. There is high-attenuation livor mortis in the sagittal sinus.

the entire body. However, asymmetric or focal gas collections should be viewed as suspicious and related to an underlying pathologic process or injury unless there is an explanation for focal or asymmetric decomposition.

Imaging Findings

Livor Mortis — Hemoconcentration from postmortem livor mortis results in increased attenuation of the affected organs, vasculature, and tissues on MDCT (Shiotani et al. 2002). This finding is easily observed in large-caliber arteries and veins as well as the cardiac chambers. In these structures, blood separates into serum and erythrocytic components due to the effect of gravity. This produces a fluid level on MDCT (Figures 3.1 and 3.2). High-attenuation erythrocytes layer dependent to plasma in the cardiac chambers and, when supine, in the posterior aspects of the great vessels (Figures 3.1 and 3.2). In the largest published study of the CT appearance of livor mortis, the gravity-dependent distribution

of intravascular erythrocytes was observed in half of the subjects imaged within 2 hours after death (Shiotani et al. 2002). This time period correlates with our knowledge of the timing of the appearance of livor mortis on gross examination of a body. The vessel wall on the nondependent side appears relatively dense because of the attenuation difference between the vessel wall and serum component of the blood (Figures 3.1 and 3.2). This is most noticeable in the aorta. Smaller-caliber blood vessels such as the cerebral arteries may also have higher attenuation (Figure 3.10c).

In our experience with postmortem MDCT, intravascular high attenuation from postmortem livor mortis is also frequently observed in the posterior dural sinuses of the cranial fossa (Figures 3.10 and 3.11). In a body that has been in the supine position after death, we commonly observe a high-attenuation posterior sagittal sinus, straight sinus, and transverse sinus. In some cases, a hematocrit effect or fluid level may be seen.

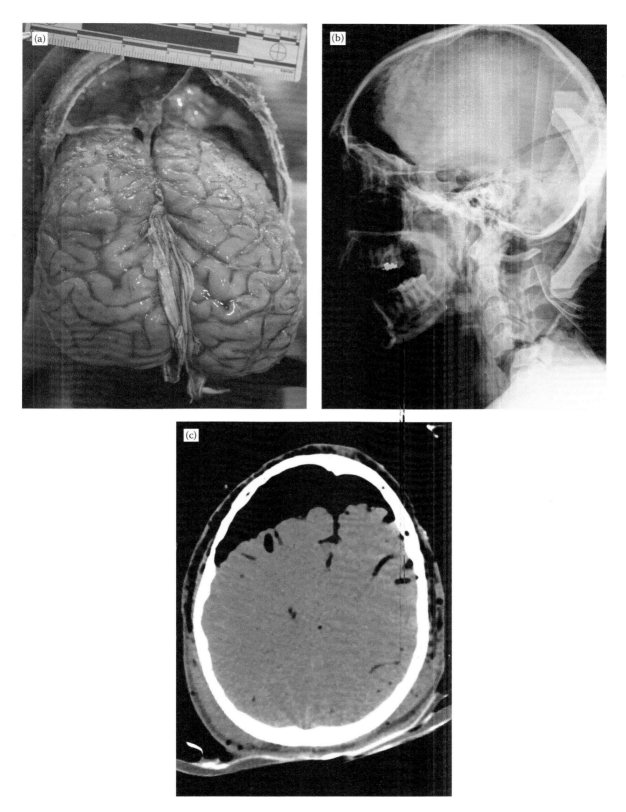

FIGURE 3.12 Moderate decompositional changes in the brain 72 hours postmortem. (a) Autopsy photograph of the brain shows softening and pinkish gray discoloration. (b) Lateral radiograph of the skull shows gas in the nondependent portion of the skull and posterior dependent settling of the brain. (c) Axial MDCT shows partial settling of the brain in the posterior, dependent portion of the calvarium with putrefactive gas filling the anterior calvarium, vasculature, and soft tissues of the scalp.

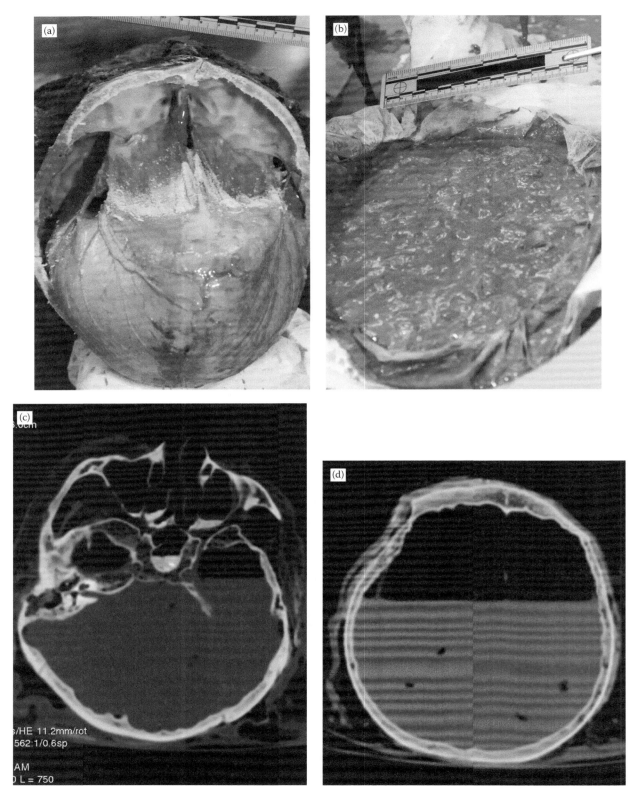

FIGURE 3.13   Advanced decompositional changes in the brain 14 days postmortem in a drowning victim. (a, b) Autopsy photograph shows liquefied brain held together by an intact dura mater in (a). When the dura is opened in (b), the liquefied brain spills out. (c, d) Axial MDCT shows liquefaction of the brain with a fluid level. Putrefactive gas is present in the calvarium and soft tissues. The scalp has peeled away from the calvarium. High attenuation sand is found in the paranasal sinuses in (c).

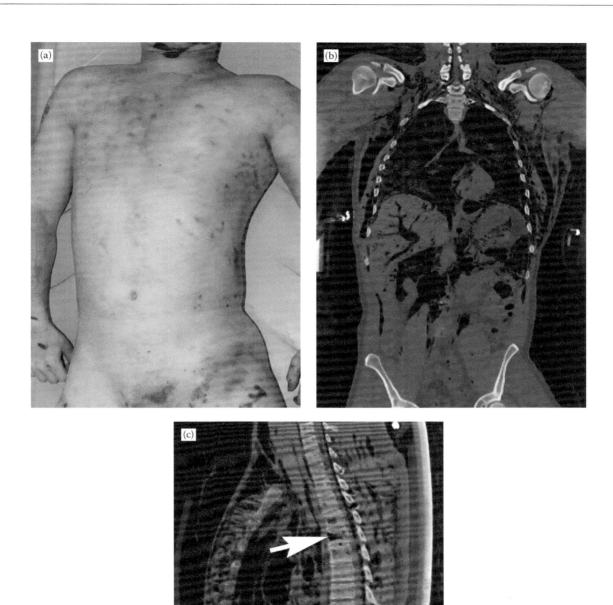

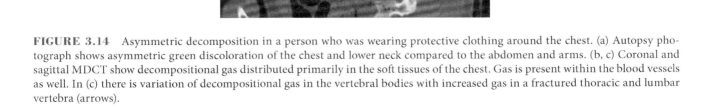

FIGURE 3.14 Asymmetric decomposition in a person who was wearing protective clothing around the chest. (a) Autopsy photograph shows asymmetric green discoloration of the chest and lower neck compared to the abdomen and arms. (b, c) Coronal and sagittal MDCT show decompositional gas distributed primarily in the soft tissues of the chest. Gas is present within the blood vessels as well. In (c) there is variation of decompositional gas in the vertebral bodies with increased gas in a fractured thoracic and lumbar vertebra (arrows).

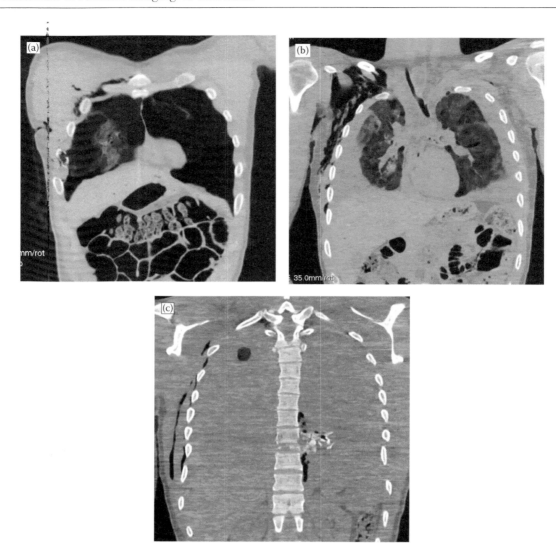

FIGURE 3.15 Asymmetric gas from a gunshot wound to the right chest with accompanying decomposition. (a, b, c) Coronal MDCT images of the chest show a focal gas collection in the right axilla and chest wall. There are right rib fractures, bilateral hemopneumothorax, and thoracic spine fractures from the bullet that passed from right to left through the chest and spine.

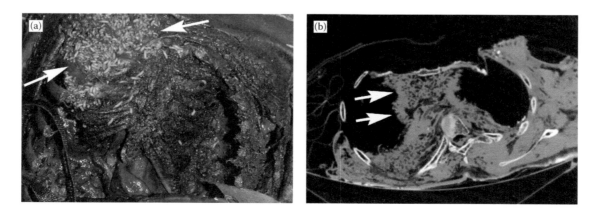

FIGURE 3.16 Advanced decomposition with partial skeletalization and maggot infestation. (a) Autopsy photograph shows maggots (arrows) and a decomposed anterior chest wall. (b) Axial MDCT shows extensive, asymmetric decomposition from blunt trauma to the thorax. All intrathoracic organs have decomposed, and there is a large cluster of maggots in the substernal region (arrows).

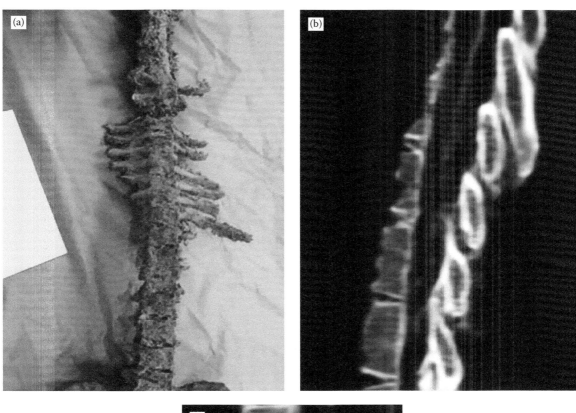

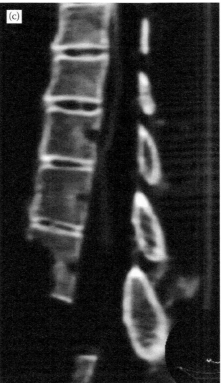

FIGURE 3.17 Animal predation on found skeletal remains. (a) Autopsy photograph shows a moth-eaten appearance to the spine, and most of the ribs have been removed. (b, c) Sagittal images of the spine show scalloping of the vertebral bodies consistent with mammalian predation.

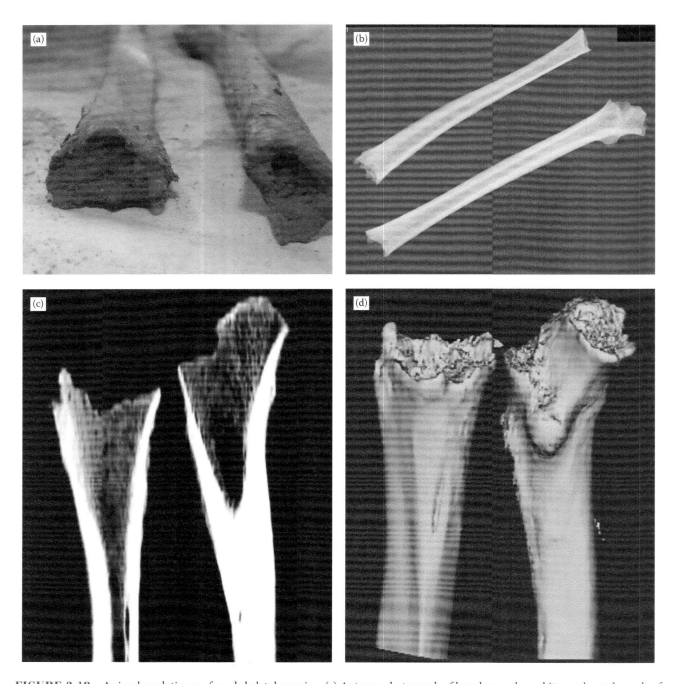

FIGURE 3.18   Animal predation on found skeletal remains. (a) Autopsy photograph of long bones shows bite marks at the ends of the bones. (b, c, d) Radiograph, coronal MDCT, and three-dimensional MDCT show scalloping of the ends of the bones consistent with mammalian bite marks.

The finding of intravascular hyperattenuation within the dural sinuses, cardiac chambers, and great vessels on postmortem MDCT should not be mistaken for thrombosis unless there are additional supportive findings. For example, cerebral or pulmonary venous thrombosis will typically have round or oval contours, may expand the involved vessels, and may be located within the nondependent portions of the vessels. In addition, there may be features of venous infarction within the brain. In pulmonary embolic disease, there may be associated enlargement of the pulmonary arteries and evidence of pulmonary infarction.

Visceral livor mortis is most commonly identified in the lung parenchyma on postmortem MDCT because of the

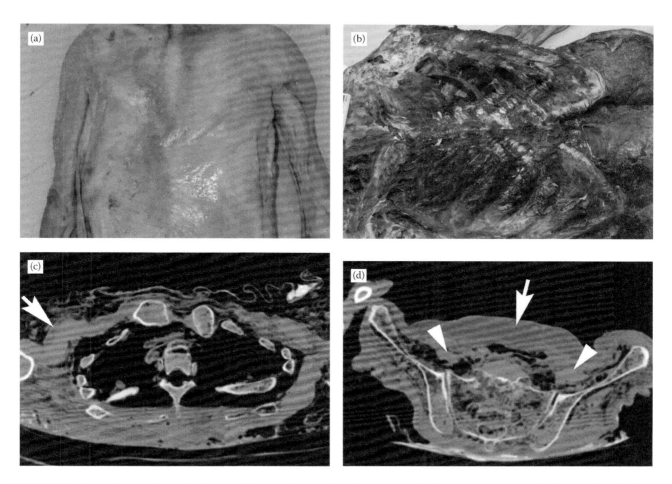

FIGURE 3.19 Adipocere formation in an adult man who was found buried in a shallow grave approximately 3 years after death. (a, b): Autopsy photograph of the unopened torso in (a) and opened chest in (b) shows brownish tan coloration of adipocere. (c, d) Axial MDCT through the upper chest and pelvis show low attenuation adipocerous subcutaneous fat (arrows), a small amount of putrefactive fluid within the chest, and the absence of lungs and mediastinal structures. The ribs are intact and slightly displaced. In the pelvis, partially decomposed organs are mixed soft tissue (arrowheads) and gas attenuation. Subcutaneous low-attenuation adipocerous fat is present.

inherent attenuation differences between aerated lung and the pulmonary vasculature. It causes an increased attenuation in the dependent lung (Figure 3.3) (Shiotani et al. 2004). There may be a vertical gradient with increasing attenuation from the nondependent to the dependent portions of the lung parenchyma with increasing degrees of livor mortis. Dependent settling of pathologic pulmonary venous congestion and edema also occurs in individuals who die from conditions that result in increased pulmonary venous pressures, such as cardiac failure, myocardial infarction, or drowning, and should not be mistaken for livor mortis. The extent and severity of the pulmonary parenchymal density are greater in pulmonary venous congestion and edema compared to normal postmortem livor mortis. Furthermore, pulmonary venous congestion and alveolar ground glass related to pulmonary

edema generally have a predominant central distribution of ground glass opacity and pulmonary venous engorgement (Levy et al. 2007). Livor mortis within the lungs may also mimic pulmonary consolidation from an infectious or neoplastic process. Careful evaluation of the distribution, contours, and presence of associated abnormalities can be helpful to differentiate livor mortis from a pathologic process.

The cutaneous and subcutaneous manifestations of livor mortis are much less profound on MDCT when compared to gross examination of the body. There is increased attenuation of the dependent subcutaneous fat and dermis. Livor mortis also causes the dermal tissues along the dependent surface of the body to be thicker when compared to the nondependent dermis (Figure 3.1).

Rigor Mortis and Algor Mortis

Postmortem MDCT shows no specific findings for rigor mortis or algor mortis. Rigor mortis does not affect the CT attenuation, size, or shape of skeletal muscles. In our practice, rigor mortis is most important in positioning the body on the CT scanner table and, in some cases, may be an obstacle to positioning and passage of the body through the CT gantry. We found that using a CT scanner that has a large gantry and a large scan field of view will accommodate the majority of bodies with rigor mortis. It is possible to physically overcome or *break* rigor. This can be considered in some cases in order to permit MDCT to be accomplished; however, this must be done in conjunction with or by the forensic pathologist because rigor status is a forensic finding and iatrogenic injury can be produced if the procedure is not done properly.

Decomposition

We found it helpful to classify the spectrum of decomposition observed on MDCT as early, moderate, and advanced (Table 3.1) (Levy et al. 2010). Even though decomposition is clearly evident at autopsy, we routinely report the degree of decomposition in our radiologic reports. This is specifically helpful when there are suspected injuries or pathologic processes that might be associated with the findings of gas in internal organs or vasculature.

One of the earliest signs of decomposition on MDCT is cerebral autolysis. There is usually some evidence of cerebral autolysis by MDCT in the majority of bodies that are

TABLE 3.1   Multidetector Computed Tomography (MDCT) Classification of Decomposition

Early	Moderate	Advanced
Cerebral autolysis	Cerebral settling	Cerebral liquefaction
Intestinal distension	Cavity fluid	Diffuse subcutaneous gas
Intestinal mural gas	Cavity gas	Diffuse visceral organ gas and organ collapse
Intravascular gas	Small amounts of subcutaneous gas	Evidence of insects or animal predation
	Small amounts of visceral gas	Skeletalization
		Adipocere formation
		Mummification

scanned and autopsied more than 24 to 48 hours after death, even if a body has been stored in the mortuary cooling chambers. The MDCT features of cerebral autolysis include blurring and loss of definition of the gray-white matter junction, decrease in cerebral attenuation, and effacement of the sulci and ventricles (Figure 3.10). During this period, the brain may exhibit mild softening at gross inspection but is otherwise macroscopically normal in appearance. Within 2 to 3 days, there is progression of autolysis and complete loss of gray-white matter differentiation on MDCT, and the cerebral ventricles and sulci become effaced (Figure 3.11). As the brain softens, it settles in the gravity-dependent portion of the calvarium, and gas fills the nondependent portion of the calvarium (Figure 3.12). At this stage, putrefactive gas may be present within the vascular structures and intracranial spaces. However, putrefactive gas may also be seen within the vasculature before brain settling is evident on MDCT. Finally, with complete cerebral liquefaction, the brain is water attenuation on MDCT, and there is a fluid level within the calvarium (Figure 3.13).

The intestinal wall and mesenteric and portal venous systems are generally the first sites of putrefactive gas on MDCT in addition to the small and large intestine, which may be distended with gas from proliferation of intestinal bacteria (Figure 3.4). Body cavities such as the pleural and peritoneal spaces may contain a small amount of fluid. The fluid may be putrefactive fluid (purge fluid) or liquefied fat. The latter is more common in the abdominal cavity from liquefied omental, mesenteric, and retroperitoneal fat. Small volumes (10 to 20 mL) of pleural fluid are considered normal at autopsy and are typically easily differentiated from pathologic collections of fluid in the pleural cavity. As putrefactive decomposition progresses, gas enters all vascular structures and potential anatomic spaces (Figure 3.5 through Figure 3.7). Putrefactive gas is normally symmetrically distributed throughout the body unless there is focal or asymmetric decomposition from an underlying injury or a focal cause of warming or cooling of the body (Figure 3.7) (Levy et al. 2010).

Although the pancreas and adrenal glands are among the earliest internal organs to undergo autolysis, they generally have a normal appearance on postmortem MDCT until putrefactive gas is present. The lack of early MDCT attenuation change in the pancreas and adrenal glands is likely due to the noncontrast techniques used in postmortem scanning. With moderate putrefactive decomposition, the pancreas may be observed to have a disproportionate amount of gas compared to other abdominal organs (Figure 3.7). Gas appears within the vasculature of the visceral organs in the early postmortem period at the same time that gas appears in other vessels throughout the body. The CT attenuation of the solid visceral organs such as the liver, spleen, and kidneys does not change until the advanced stages of decomposition when the organs begin to fragment, degenerate, and liquefy. Rounded lucencies filled with putrefactive gas will eventually appear within the organs surrounded by intact connective tissues (Figure 3.5 and Figure 3.8). Therefore, even in moderate stages of decomposition, visceral organs are still normal in shape and contour. Eventually, the connective tissues collapses, and the organs are not recognizable in shape and appearance (Figure 3.9). Gas fills the abdominal and chest cavities as the organs collapse and liquefy (Knight and Saukko 2004).

Insect activity and animal predation are important diagnostic features of decomposition because they provide clues as to the location of a body at the time of death and information that may assist in determining if a body has been moved after death. The features are usually easily recognizable on gross examination, and postmortem imaging does not assist the forensic pathologist with these findings. Most importantly, the appearance of insect activity and animal predation should not be mistaken for pathologic processes or injury when interpreting postmortem MDCT. Larvae are attracted to moist areas of the body, such as the mucous membranes of the nasal and oral cavities or sites of external injury. Consequently, they are commonly observed in the facial and head regions as well as at sites of injury. Larvae appear as linear and curvilinear soft tissue or surface irregularities on MDCT (Figure 3.16). Animal predation is characterized by one or more bite marks in soft tissues and bones. Often there are innumerable marks from small mammals such as rodents or dogs gnawing on the corpse (Figures 3.17 and 3.18).

The formation of adipocere in severely decomposed bodies preserves the subcutaneous tissues and portions of internal organs. It has a characteristic low-attenuation appearance on MDCT. Because adipocere formation frequently coexists with putrefactive decomposition, both processes may be present on MDCT, revealing a body that has absent or partially decomposed internal organs and intact skeletal structures surrounded by adipocerous subcutaneous fat (Figure 3.19).

CONCLUSIONS

Postmortem change and decomposition should be considered normal findings on postmortem MDCT. Thorough knowledge of the MDCT findings of postmortem change and the stages of decomposition are important when interpreting postmortem MDCT, because these processes may be mistaken for pathologic processes or injury. They may also make interpretation of pathologic processes that are relevant to the cause of death more difficult.

REFERENCES

Ambrosi, L., and Carriero, F. 1965. Hypostasis in the internal organs. Histological aspects. *J Forensic Med* 12: 8–13.

Di Maio, V. J. M., and Di Maio, D. J. 2001. *Forensic pathology,* Boca Raton, FL: CRC Press.

Dix, J., and Graham, M. A. 2000. *Time of death, decomposition, and identification: an atlas,* Boca Raton, FL: CRC Press.

Knight, B., and Saukko, P. J. 2004. *Knight's forensic pathology,* London: Oxford University Press.

Levy, A. D., Harcke, H. T., Getz, J. M. et al. 2007. Virtual autopsy: two- and three-dimensional multidetector CT findings in drowning with autopsy comparison. *Radiology* 243: 862–868.

Levy, A. D., Harcke, H. T., and Mallak, C. T. 2010. Postmortem imaging: MDCT features of postmortem change and decomposition. *Am J Forensic Med Pathol* 31:12–17.

Mann, R. W., Bass, W. M., and Meadows, L. 1990. Time since death and decomposition of the human body: variables and observations in case and experimental field studies. *J Forensic Sci* 35: 103–111.

Payne-James, J., Busuttil, A., and Smock, W. (Eds.). 2003. *Forensic medicine: clinical and pathological aspects,* San Francisco: Greenwich Medical Media.

Shiotani, S., Kohno, M., Ohashi, N., Yamazaki, K., and Itai, Y. 2002. Postmortem intravascular high-density fluid level (hypostasis): CT findings. *J Comput Assist Tomogr* 26: 892–893.

Shiotani, S., Kohno, M., Ohashi, N. et al. 2004. Non-traumatic postmortem computed tomographic (PMCT) findings of the lung. *Forensic Sci Int* 139: 39–48.

Chapter 4

Gunshot Wounds

FORENSIC PRINCIPLES

Deaths from gunshot wounds are an all too common part of daily life in the United States. Annually, 30,000 persons per year, approximately 80 persons per day, die from gunshot wounds inflicted as a result of homicide, suicide, and accident (Centers for Disease Control and Prevention 2008). Medical examiners in many states are legally mandated to investigate gunshot wound deaths and commonly use radiologic methods to aid forensic autopsy of gunshot wound victims.

Radiography was first applied to the investigation of gunshot wounds almost immediately after the discovery of X-rays in 1895 (Brogdon 1998). It is used to locate the bullet, identify the class of ammunition and possibly the weapon used, assist in the documentation of the path of the bullet, and assist in the retrieval of the bullet (Di Maio 1999). The National Association of Medical Examiners recommends that radiographs be obtained as part of the investigation of any gunshot wound fatality and that recovery and documentation of foreign bodies be made for evidentiary purposes (Peterson and Clark 2006).

Ballistics

Ballistics is the study of projectile motion. Internal ballistics is the study of the bullet within the firearm, external ballistics examines the motion of the bullet in flight, and terminal ballistics evaluates the bullet in the target. The latter includes wound ballistics, which is the interaction of bullets within the human body (Wilson 1999).

Three types of guns are used in the civilian population: handguns, rifles, and shotguns. Handguns and rifles are categorized by the size of the ammunition fired (caliber or millimeters), operating characteristics (lever, pump, or bolt action), and firing rate (single shot, semiautomatic, or automatic). They both fire ammunition similar in construction, with a cartridge element that consists of a case, primer, and propellant capped by a bullet. Bullets are sized by diameter, which is expressed as a decimal fraction of an inch or in millimeters. The term *caliber* refers to the internal diameter of the gun barrel or the diameter of the

bullet. Bullets are usually made of lead. They may contain a core element and a covering element. The covering element is commonly referred to as the *jacket*. The jacket is usually a different metal than the bullet. It is most commonly copper or a copper alloy that partially or fully covers the lead bullet (Figures 4.1 and 4.2). This is important to recognize, because the jacket and bullet may separate and appear quite different on radiography and multidetector computed tomography (MDCT) imaging (Figure 4.2). Both can deform and fragment depending on their composition and construction, the intermediate targets they pass through or strike, and their interactions with specific types of body tissues (Figures 4.2 and 4.3).

Shotguns differ from rifles and handguns because they have smooth bores, whereas the other two have grooves in their barrels. Grooves impart a rotational spin to the bullet that increases stability during flight and therefore improves accuracy. Rather than a bullet, a shotgun uses a cartridge, which is a paper or plastic tube with a metal base, primer, and propellant. Metal spheres, called *pellets* or *shot,* are contained within most cartridges, but there are solid projectiles, called *slugs,* in some types of shotgun ammunition. The pellets vary in size and number with the type of cartridge (Figure 4.4). Shotgun bore diameters are usually referred to by gauge instead of caliber. The gauge of a shotgun (except for .410) equals the number of solid spheres of lead that each has the same diameter as the inside of the gun barrel which would weigh 1 pound. Therefore, a small-gauge shotgun has a larger barrel diameter than a large gauge shotgun. For example, a 20-gauge shotgun has a smaller diameter than a 12-gauge shotgun. When a shotgun is fired, the pellet distribution generally spreads outward at a distance, which is controlled by the construction of the end of the barrel, called the *choke.* The shot distribution pattern is also influenced by collisions of pellets within the tissue. Consequently, systems for determining range in shotgun wound cases may be relatively inaccurate (Brogdon 1998).

Tissue damage is determined by the amount of kinetic energy imparted to surrounding tissue by the bullet. Bullet mass and velocity determine the kinetic energy at the time

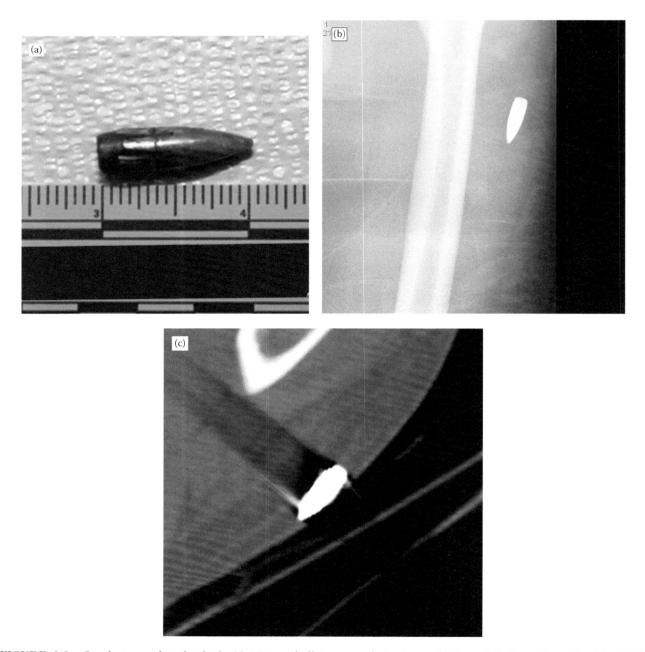

FIGURE 4.1 Gunshot wound to the thigh. (a) 7.62 mm bullet recovered at autopsy. (b) Frontal digital radiograph of the thigh shows the bullet in the soft tissue of the lateral left upper thigh. The shape and configuration of the bullet on the radiograph are identical to the recovered bullet. (c) Sagittal oblique MDCT in the plane of the long axis of the bullet shows metallic streak artifact and blurred edges.

of impact. Of these two factors, velocity is usually the most important. Bullet mass varies by ammunition construction, and velocity varies by weapon type. The longer the weapon barrel, usually the higher the bullet velocity, which is measured in feet or meters per second. Most rifles have a higher muzzle velocity than handguns and shotguns. As a general rule, high-velocity weapons will cause more tissue damage than low-velocity weapons. Other factors that contribute to the degree of tissue damage include kinetic energy loss as a function of bullet shape, tumbling (yaw), and the density of the interacting tissue.

There are three mechanisms of tissue damage from bullets: laceration (or crushing), shock wave, and cavitation.

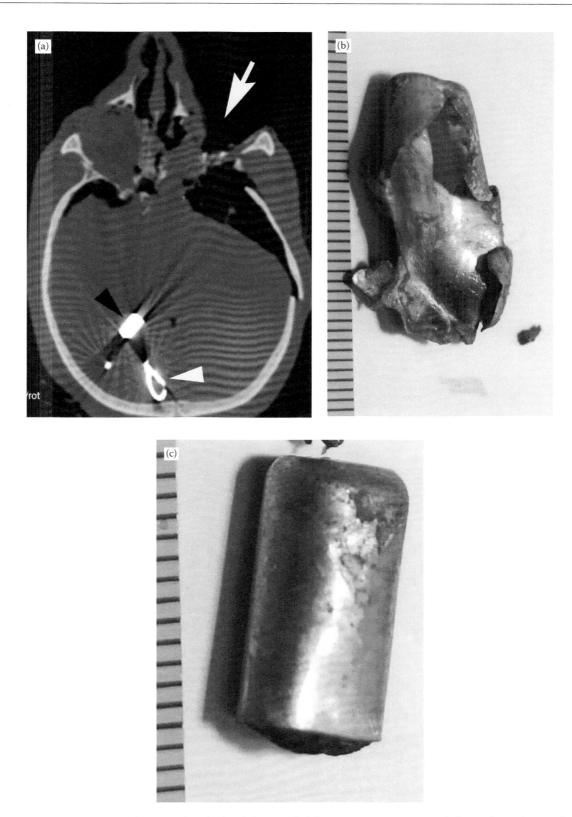

FIGURE 4.2   Penetrating gunshot wound to the head that traveled from anterior to posterior, left to right, and upward. (a) Axial MDCT shows the bullet entered the left orbit (arrow). There are multiple orbital and nasal fractures, pneumocephalus, and calvarial fractures. Three bullet components are lodged in the right posterior brain. The bullet jacket (white arrowhead) and a tiny lead fragment are separate from the penetrator (black arrowhead). (b) Copper jacket recovered at autopsy. (c) Penetrator recovered at autopsy.

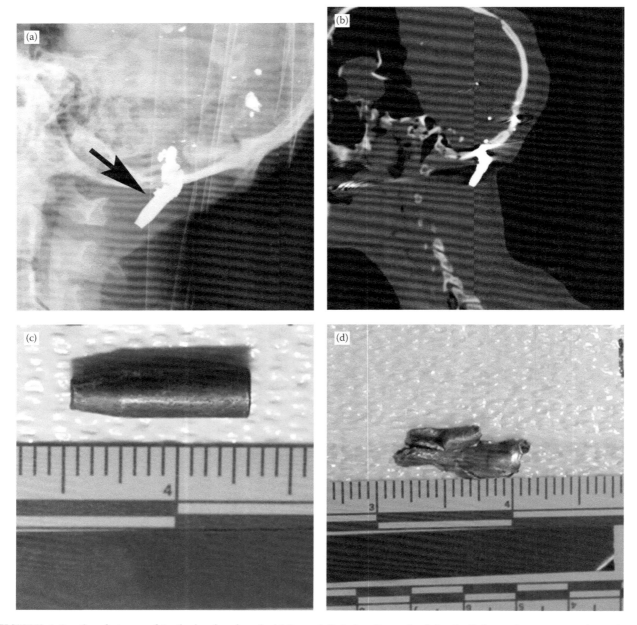

FIGURE 4.3 Gunshot wound to the head and neck. (a) Lateral digital radiograph of the skull shows the penetrator (arrow) and bullet fragments. (b) Sagittal MDCT shows the penetrator and bullet fragments with metal streak artifact and blurred edge detail. (c) Penetrator and (d) jacket fragment recovered at autopsy.

Initially, as a bullet passes through tissue, it creates an expanded void region or temporary cavity. This temporary cavity collapses in a few milliseconds, leaving a second type of cavity—a smaller permanent cavity or wound track. When a bullet has an unusual entrance wound pattern, different tissue effect, or course from its predicted ballistics, one should consider that the bullet passed through an intermediary target. If the intermediary target is of sufficient mass and resistance, the bullet can destabilize

or break up and more readily lose kinetic energy when it strikes the victim. This may increase wound severity and alter the pattern of bullet fragmentation as the projectile expends more kinetic energy within the tissue rather than passing through the tissue (Di Maio 1999).

Autopsy Findings

Autopsy of gunshot wound victims begins with the identification of all wounds on the external surface of the body

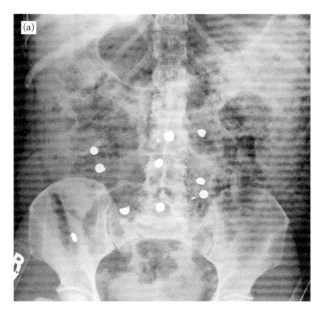

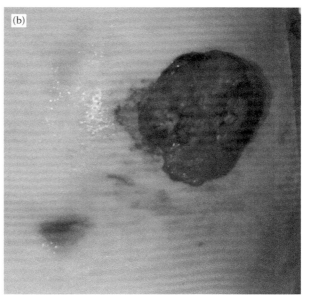

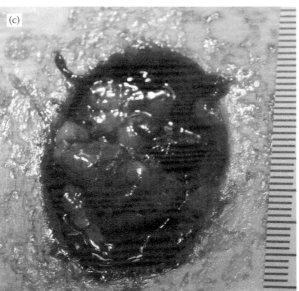

FIGURE 4.4 Shotgun wound of the abdomen. (a) Digital abdominal radiograph shows nine ovoid metal pellets, one smaller irregular pellet, and tiny metallic fragments. The pellets are grouped together, indicating the shotgun was positioned at close range but did not have contact with the skin. (b, c) Autopsy photographs of the anterior abdominal wall show a single large entrance wound that measures 3.5 cm by 2.5 cm. The wound does not have soot or stippling on the skin. The green discoloration along the medial aspect of the wound is from early decomposition.

to determine if they are entry or exit wounds. Attempts are made to estimate the distance between the weapon muzzle and the body and the angle at which the bullet entered the body. Medical examiners use a well-defined taxonomy for gunshot wounds that should also be applied to radiologic findings. However, some of the taxonomy cannot be applied to radiologic studies because some features of gunshot wounds can only be identified by direct observation. For

example, skin surface features such as wound shape, pigmentation, discoloration, and soot deposition are findings that can only be made on external examination of the body.

The distance, or range, between the weapon muzzle and the body surface is judged to be contact, near contact, intermediate, or distant (or indeterminate) when examining the entrance wound at autopsy. Contact gunshot

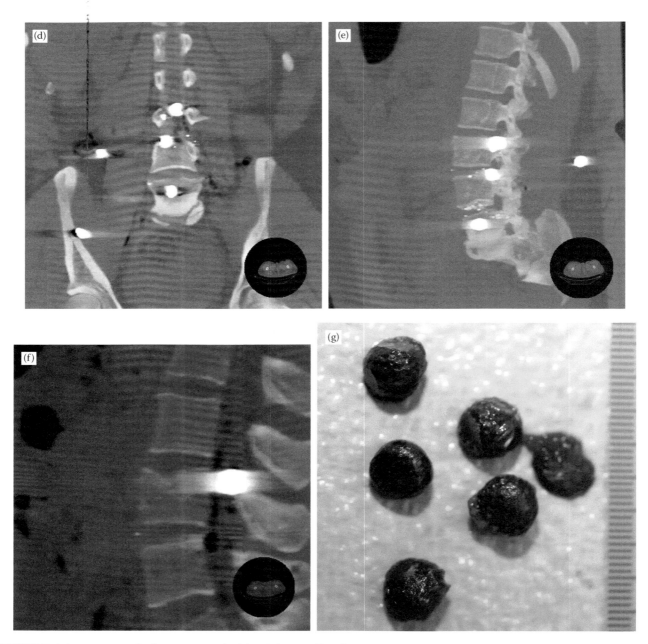

FIGURE 4.4 (*Continued*) Shotgun wound of the abdomen. (d, e, f) Coronal and sagittal maximum intensity projection MDCT shows that two of the pellets are embedded in the lumbar spine and one is in the spinal canal. (g) Recovered pellets are approximately 9 mm in diameter (.32 caliber), consistent with double-O buckshot. A shotgun shell containing double-O buckshot typically contains nine pellets. The single fragmented pellet accounts for the irregular pellet and tiny metallic fragments identified on the radiograph.

wounds are classified as hard contact or loose contact wounds. Hard contact and loose contact wounds are differentiated by the pattern of tissue searing from the hot gasses and soot emitted from the muzzle of the gun (Figure 4.5a). The appearance of the wound margin allows the further determination of contact angle or incomplete contact for close-range wounds. Searing and soot patterns on the skin surface may rarely suggest a specific weapon. Intermediate-range wounds show tiny, punctate dermal abrasions produced by unburned particles of gunpowder

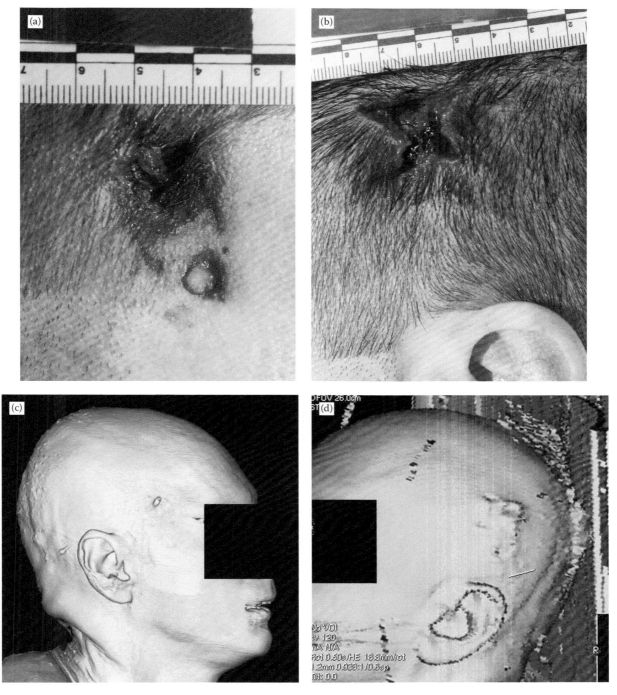

FIGURE 4.5 Hard-contact, perforating gunshot wound of the head that entered the skull from an anterior to posterior, right to left, and upward direction. (a) Autopsy photograph shows the entrance wound with marginal abrasion and soot. Note that the outline of the gun muzzle, a 9mm pistol, has seared the skin, and there is an oval mark below the entry wound from the operating rod. (b) Autopsy photograph of the exit wound shows a stellate pattern. (c, d) Three-dimensional MDCT reconstructions using surface algorithm.

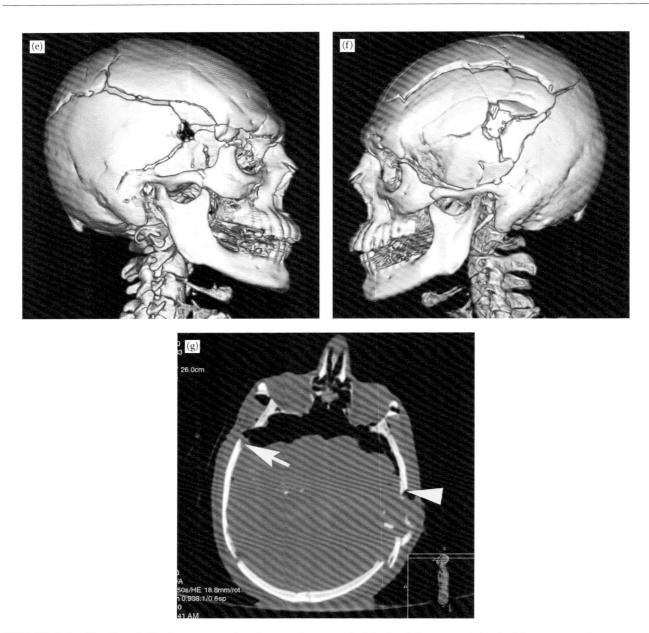

FIGURE 4.5 (*Continued*) Hard-contact, perforating gunshot wound of the head that entered the skull from an anterior to posterior, right to left, and upward direction. (e) Three-dimensional MDCT of the skull shows the right entry wound and calvarial fractures. (f) Three-dimensional MDCT of the skull on the left shows the exit wound and associated fractures. (g) Axial MDCT in bone window setting shows a right temporal fracture defect with inward beveling (arrow) indicating the entry of the bullet and a larger left temporal defect that has an external bevel (arrowhead) indicating the bullet's exit.

striking the skin. This is often referred to as *stippling* and indicates that the weapon was close but not in contact with the body (Denton et al. 2006). Stippling can also be seen on adjacent body parts, such as the hand in self-inflicted gunshot wounds if the hand is near the weapon muzzle and there is loose skin contact (Figure 4.6). Unlike soot, stippling cannot be washed away. Soot and burn patterns

are also indicative of the weapon type and muzzle adaptation. These patterns are related to diversion of hot gasses from ports, vents, and compensators at the muzzle. Distance gunshot wounds occur far enough from the body that neither soot nor stippling occur. The distance at which this occurs depends on the gun type. In general, no soot or stippling occurs with a handgun fired at a distance

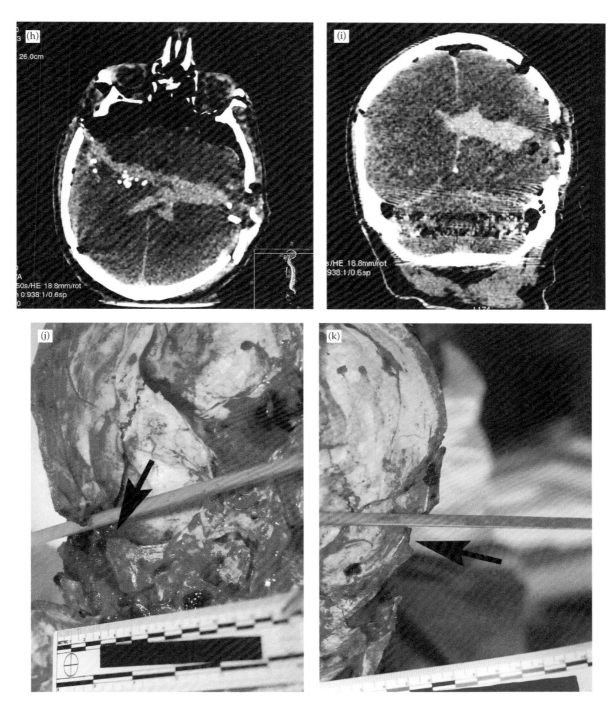

FIGURE 4.5 (*Continued*) Hard-contact, perforating gunshot wound of the head that entered the skull from an anterior to posterior, right to left, and upward direction. (h, i) Axial and coronal MDCT with soft tissue window setting shows hemorrhage in the wound as high attenuation. The bone and tiny metal fragments in the brain indicate a right-to-left direction of the bullet path. Postmortem-dependent settling of the brain moves the wound track posterior. The entry and exit points in the bone determine the true linear alignment. (j, k) Autopsy photographs show internal beveling of the skull at the entry site (arrow in j) and external beveling at the exit site (arrows in k).

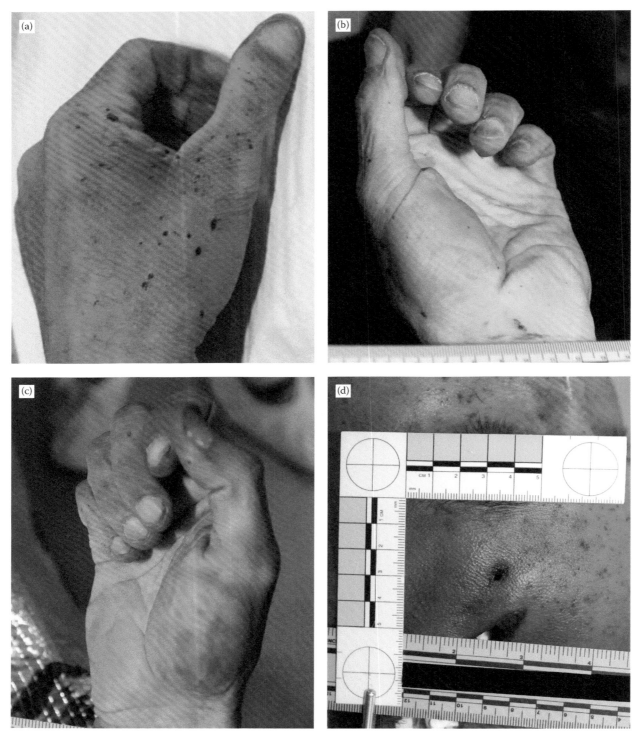

FIGURE 4.6   Stippling on the hands (a, b, c) and face (d) in cases of self-inflicted gunshot wounds. Gunpowder stippling on the hand indicates that the body part was near the muzzle, and the muzzle was in loose contact.

of 2 feet and a rifle fired at a distance of 3 feet. Beyond this critical distance, it is not possible to judge the range at which firing occurred (Di Maio 1999).

An abrasion ring, which can be concentric or eccentric, characterizes most entrance wounds. Bullets that enter the body perpendicular create a concentric entry wound; those that enter at an angle create an eccentric entry wound. Bullet characteristics and the area of the body that is struck also influence abrasion ring configurations. Entrance gunshot wounds over bony surfaces such as the skull may have a stellate pattern rather than an abrasion ring. In shotgun wounds, the muzzle-to-body distance affects the number and size of wounds. At close range, the shotgun pellets are closely grouped and produce a single large entry wound (Figure 4.4b). The size of the entrance wound cannot be used to determine bullet caliber, because skin elasticity, folds, and stretching affect the size of the wound. Multiple small wounds from the spray of individual shotgun pellets characterize a distant shotgun wound. The appearance of entrance wounds over bone will not exactly match the bullet caliber because the size of the hole is determined by bullet construction, ballistic properties, and bullet diameter (Di Maio 1999).

Exit wounds are typically larger and more irregular than entrance wounds (Figures 4.7a,b). This is attributed to loss of bullet spin stability (yaw) or the deformity of the bullet as it passes through body tissue (Di Maio 1999). The exit wound can have *shoring,* which is a broad, irregular margin caused by reinforcement of the skin by contact with a firm surface or tight clothing. The anatomic area of exit also affects the size and configuration of the exit wound. If the bullet exits in an area of lax skin, the wound is usually small or slit like. In contrast, if the bullet exits through tightly stretched skin, the exit wound it is likely to be more round or stellate.

Gunshot wounds are further classified as penetrating or perforating. Bullets that enter the body but do not exit are penetrating gunshot wounds. The bullet remains in the body as a single fragment or multiple fragments, depending on the bullet material and the interaction of the bullet with intermediate targets such as bone (Figure 4.2). Bullets that enter and then exit the body are perforating gunshot wounds. If the bullet remains intact, there will be no residual metallic fragments in the body (Figures 4.8 and 4.9). If the bullet fragments along its course, metallic fragments will be deposited within the tissue along the course of the gunshot wound track (Figure 4.5h). Gunshot wounds can be a combination of penetrating and perforating. For example, a gunshot wound to the chest where the bullet passes through the heart and lodges in the left lung is a perforating wound of the heart and penetrating wound of the chest. A bullet may also fragment with portions remaining in the body and others exiting the body. A gunshot wound track or path is determined from the location of the entrance and exit wound and the direction taken by the bullet. The bullet path is customarily described in three directions that define the orientation or vector of travel. The description is made in relation to the standard anatomic position and may not represent the actual pathway vectors at the time of the gunshot wounds. The description indicates whether the bullet passes through the body from *anterior to posterior* or *posterior to anterior, left to right* or *right to left,* and *upward* or *downward.*

The determination of the direction for perforating gunshot wounds relies on several other features in addition to the characterization of entry and exit wounds. One of the most helpful features is the passage of the bullet through bone. When a bullet enters and exits bone, marginal fracturing creates beveling, with the bevel occurring outward in the direction of travel (Figure 4.5g and Figure 4.10a). In addition, the bone fragments created by the perforating bullet are propelled laterally and forward in the direction of the bullet. There are instances of exception to virtually every dictum, however, because bone varies in thickness, contour, and composition (Figure 4.11). Very thin bones may not show directional beveling, and the flat bones of the skull show variations in beveling at the inner and outer tables depending on angle of incidence and range (Coe 1982, Di Maio 1999). Metallic fragments in the tissues of the wound track are also an indicator of directionality, because they are usually carried forward along the direction of projectile travel. If a bullet fragments, pieces of metal will be distributed along the bullet track and may commingle with bone fragments that are also deposited along the bullet track (Figure 4.10a) (Harcke et al. 2008).

In general, a bullet entering and passing through the body maintains trajectory without deviation unless it strikes bone and loses velocity. When this occurs, the bullet's path may deviate or change direction. Bullets may also fragment, with the fragmented portions deviating from the primary wound track (Figure 4.12). Furthermore,

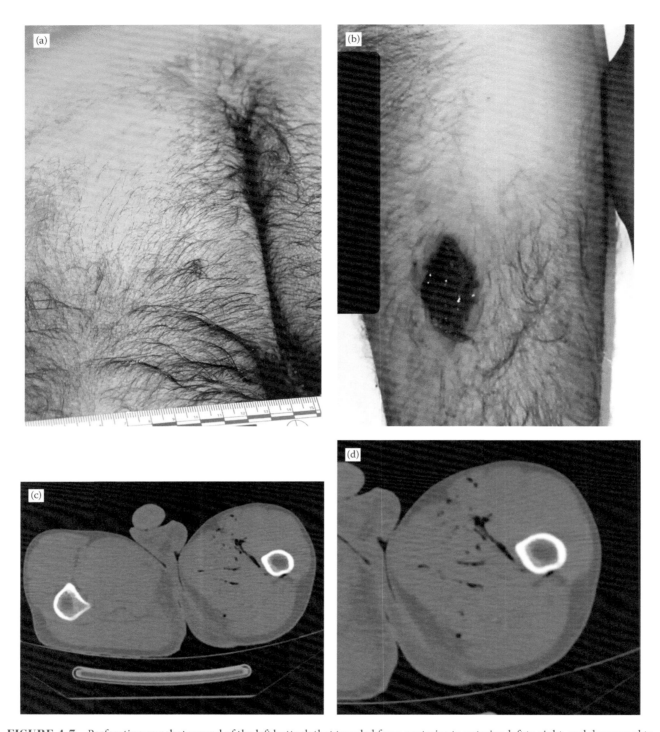

FIGURE 4.7   Perforating gunshot wound of the left buttock that traveled from posterior to anterior, left to right, and downward to lacerate the left femoral artery. (a) Autopsy photograph shows the entrance wound on the left buttock. (b) Autopsy photograph shows the exit wound on the anterior left thigh. (c) Axial MDCT shows soft tissue air in the left leg. (d) Axial MDCT at entrance level shows the wound obscured but tiny air collections in the track.

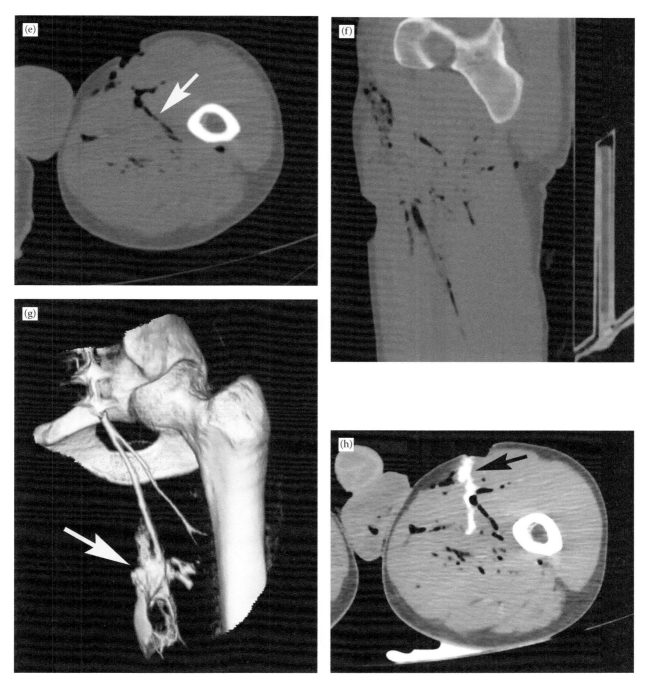

FIGURE 4.7 (*Continued*) Perforating gunshot wound of the left buttock that traveled from posterior to anterior, left to right, and downward to lacerate the left femoral artery. (e) Axial MDCT at the level of the exit wound shows a linear air collection in the soft tissue consistent with the wound track (arrow). (f) Oblique sagittal MDCT in track, entrance obscured, exit visible. (g) Three-dimensional MDCT angiography shows the common femoral, proximal superficial femoral, and profunda femoris arteries. Intravenous contrast extravasates (arrow) from laceration of the proximal superficial femoral artery. (h) Axial MDCT at the level of extravasation during the angiogram shows the superficial femoral artery laceration and contrast leakage. The exit wound is also visible on the anterior thigh.

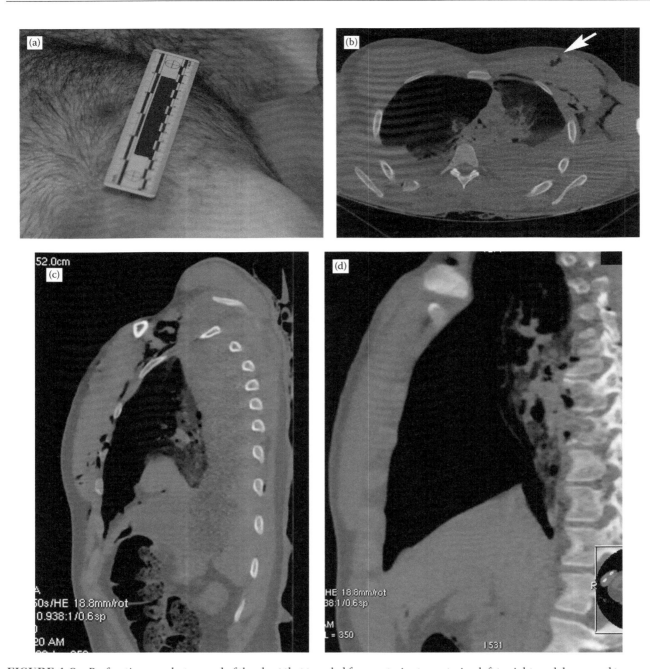

FIGURE 4.8   Perforating gunshot wound of the chest that traveled from anterior to posterior, left to right, and downward to cause a bilateral pneumohemothoraxes. (a) Autopsy photograph shows the entrance wound above the left nipple. (b) Axial MDCT at the level of the entrance wound shows a gas collection in the pectoral muscle at the site of the entrance wound (arrow) as well as gas dissecting through the left chest wall. There is mediastinal hemorrhage and bilateral pneumohemothoraxes. Note that the mediastinum is shifted to the left from the tension pneumohemothorax on the right. (c, d, e) Sagittal MDCT images show skeletal fractures reflecting the path of the bullet—anterior fourth left rib, seventh and eighth vertebral bodies, fifth and sixth right posterior ribs, and the right scapula.

analysis of ballistic injury should always consider the position of the body at the time of injury, especially when entry and exit wounds do not match other points on the track, such as fractures. This should raise suspicion that postmortem position is significantly different from the position of the body at the time of injury (Figure 4.8 and Figure 4.13). Another occurrence that can confound determination of a wound track is migration of the bullet within the body which then becomes an embolus. This is of particular concern when the bullet settles in a body

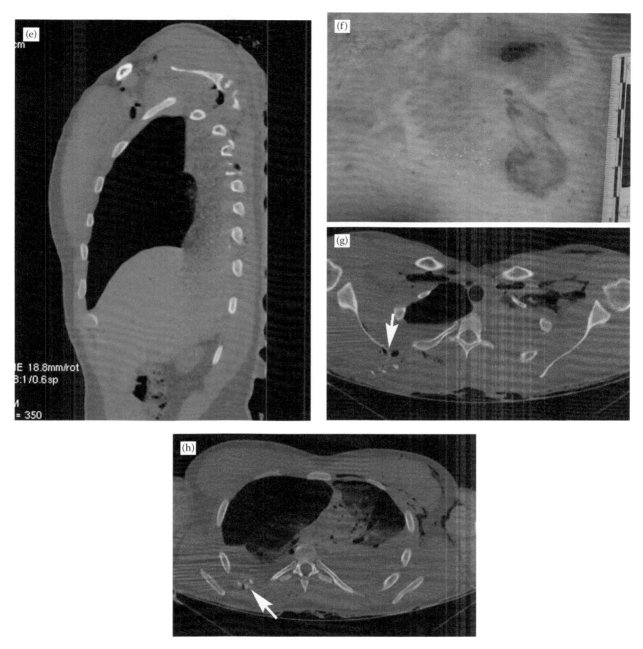

FIGURE 4.8 (*Continued*) Perforating gunshot wound of the chest that traveled from anterior to posterior, left to right, and downward to cause a bilateral pneumohemothoraxes. (c, d, e) Sagittal MDCT images show skeletal fractures reflecting the path of the bullet—anterior fourth left rib, seventh and eighth vertebral bodies, fifth and sixth right posterior ribs, and the right scapula. (f) Autopsy photograph shows the exit wound in upper right back adjacent to a skin abrasion. (g) Axial MDCT at the level of the exit wound shows a marginal fracture of the medial border of the right scapula (arrow). Bone fragments from the scapula are directed posterior. No rib fracture is at this level. (h) Axial MDCT shows posterior right rib fracture (arrow) inferior to the level of the scapula fracture.

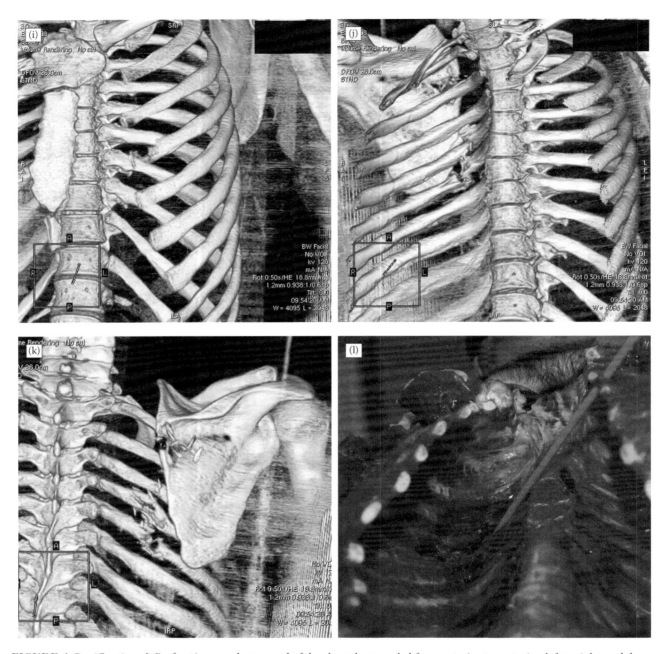

FIGURE 4.8 (*Continued*) Perforating gunshot wound of the chest that traveled from anterior to posterior, left to right, and downward to cause a bilateral pneumohemothoraxes. (i, j, k) Three-dimensional MDCT shows that the left rib fracture, spine, and right rib fractures follow a linear path. However, the scapula fractures are not in alignment. (l) Autopsy photograph with rod approximating alignment of the rib and spine fractures. The fractured vertebral elements have been resected.

cavity or tubular structure within the body, such as the vasculature, trachea and bronchi, neural canal, and urinary and gastrointestinal tract. Bullets may travel to a location far from the wound path after entering a lumen or cavity. A variety of mechanisms, such as vascular flow, peristalsis, or simply gravity, may affect migration (Figure 4.12) (Brogdon 1998).

RADIOLOGIC PRINCIPLES

Full-body radiography is used with forensic autopsy in most gunshot wound victims to document and locate all of the bullet fragments. If only frontal radiographic views are obtained, precise localization of metallic bullet fragments is limited to two planes. Orthogonal radiographic projections (frontal and lateral views) are the

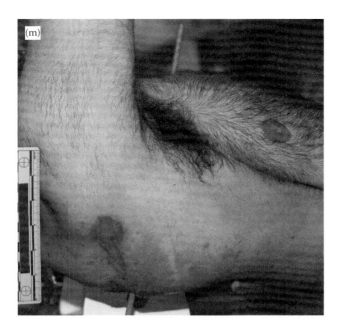

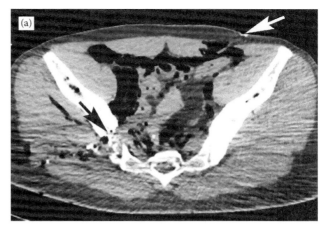

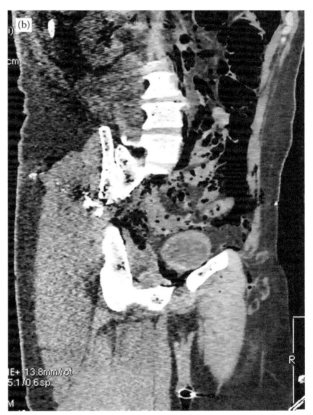

FIGURE 4.8 (*Continued*) Perforating gunshot wound of the chest that traveled from anterior to posterior, left to right, and downward to cause a bilateral pneumohemothoraxes. (m) Autopsy photograph with right arm raised (causing rotation of the scapula) shows critical points of the wound track can be brought into alignment. This reflects the position of the body at the time of shooting.

most optimal method of precise localization if radiography alone is being used. C-arm fluoroscopy may augment localization if a lateral view is not obtained. The application of MDCT to postmortem imaging has made it possible to obtain precise three-dimensional localization of bullet fragments. It has been shown that this technique is an effective method not only for localization of bullet fragments but also for documenting the gunshot wound track and evaluating internal organ injury prior to autopsy (Harcke et al. 2007, Levy et al. 2006, Thali et al. 2003).

The presence of a bullet is easily determined on radiography and MDCT by its characteristic shape and high radiographic attenuation. Measurement of its dimensions on radiography and MDCT may be limited by inherent inaccuracies caused by geometric and physical factors. Radiographs show the borders of a metallic object very clearly, but magnification always occurs because of the distance between the X-ray source and the body and the bullet and the X-ray detector. The shape and length of the object can be distorted when the object

FIGURE 4.9 Perforating gunshot wound of the abdomen that traveled from anterior to posterior, left to right. The victim was standing upright at the time of the shooting, and the bullet was shot level with respect to the ground. The apparent nonlinear tract, which appears to be diverted laterally upward after exiting the pelvis on MDCT, is the result of the supine position of the body during the scan. (a) Axial MDCT shows a fracture defect in the right sacroiliac joint (black arrow) with bone fragments distributed posterior, indicating a posterior exit. The entry wound (white arrow) is a skin defect with a high-attenuation focus on the anterior abdominal wall. Free intraperitoneal air is also present. (b) Oblique sagittal MDCT shows the plane of the wound track.

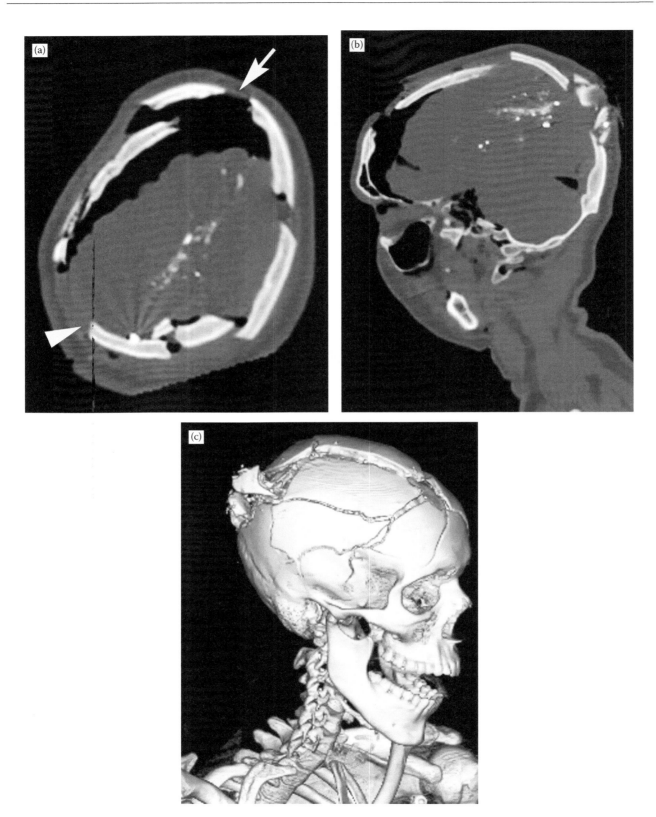

FIGURE 4.10   Perforating gunshot wound of the head that traveled from anterior to posterior, left to right, and downward. (a, b) Axial and oblique sagittal MDCT show complex calvarial fractures with internal beveling at the frontal entrance wound (arrow) and outward beveling at the occipital exit wound (arrowhead). (c) Three-dimensional MDCT shows entry and exit wounds with associated fractures.

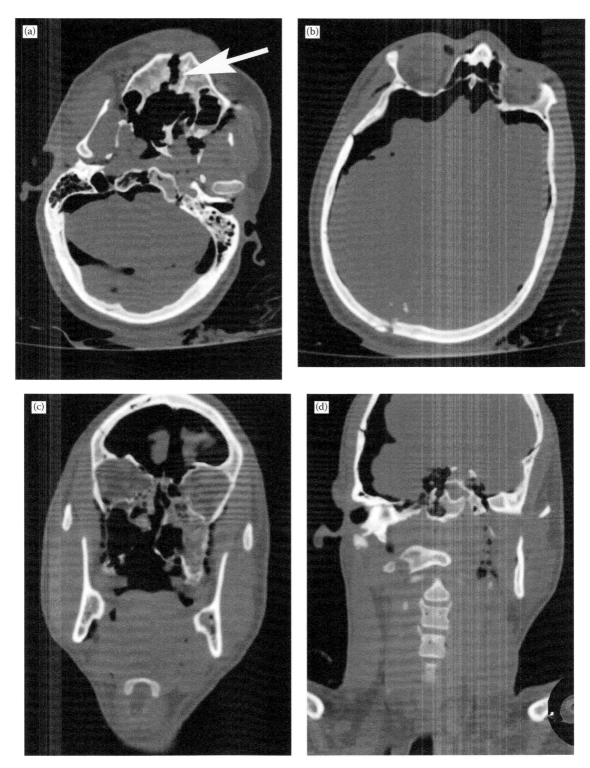

FIGURE 4.11   Suicide by intraoral gunshot. (a) Axial MDCT shows multiple maxillary fractures (arrow) and a fracture defect in the clivus. (b) Axial MDCT at the exit wound shows a fracture defect in the occiput with small bone fragments in the adjacent brain. This demonstrates an exception to the principle that fragments show the direction of the bullet. (c, d) Coronal MDCT shows the nasopharyngeal soft tissue and bone destruction and the track through the sphenoid and clivus.

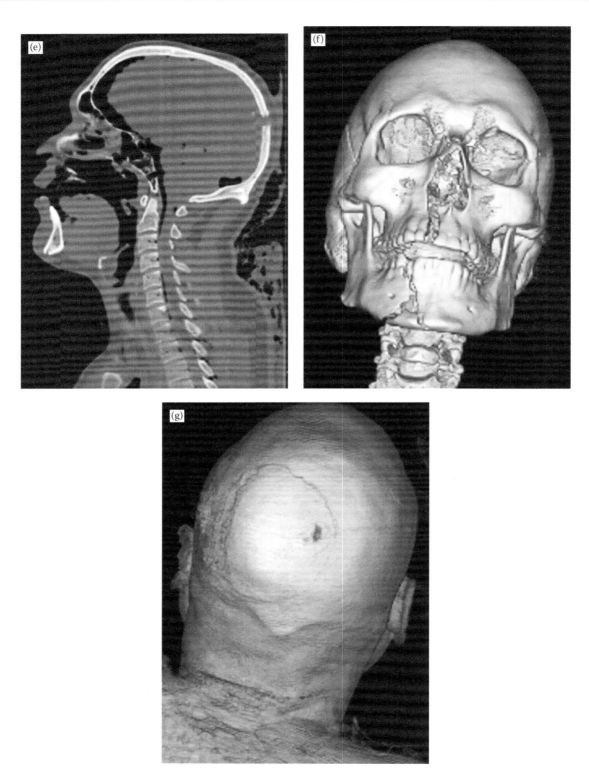

FIGURE 4.11   (*Continued*) Suicide by intraoral gunshot. (e) Oblique sagittal MDCT in the plane of the wound track. (f, g) Three-dimensional MDCT bone (f) and soft tissue (g) reconstructions clarify maxilla and mandible fractures and the exit wound.

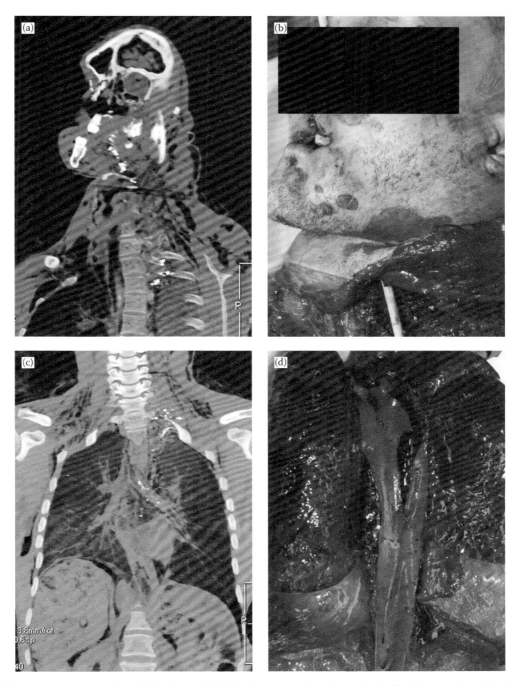

FIGURE 4.12 Penetrating gunshot of the face with bullet fragmentation shows that bullet fragment location does not always predict the course of the wound track. The bullet entered through the right maxillary sinus and traveled from anterior to posterior, right to left, and downward through the neck. (a) Oblique coronal MDCT in the plane of the wound track shows metallic fragments in the wound track at the left lung apex with associated rib fractures. (b) Autopsy photograph shows a rod in the wound track. (c) Coronal MDCT of the chest at the level of the carina shows metallic fragments in the mediastinum. Ill-defined borders of the bronchi and vascular structures suggest the presence of hemorrhage, but no specific injury is identified. Hemorrhage and metallic fragments are present in the left main stem bronchus and mediastinum. A focal hematoma is also present in the left lung apex adjacent to bullet fragments. (d) Autopsy photograph shows laceration of the descending aorta.

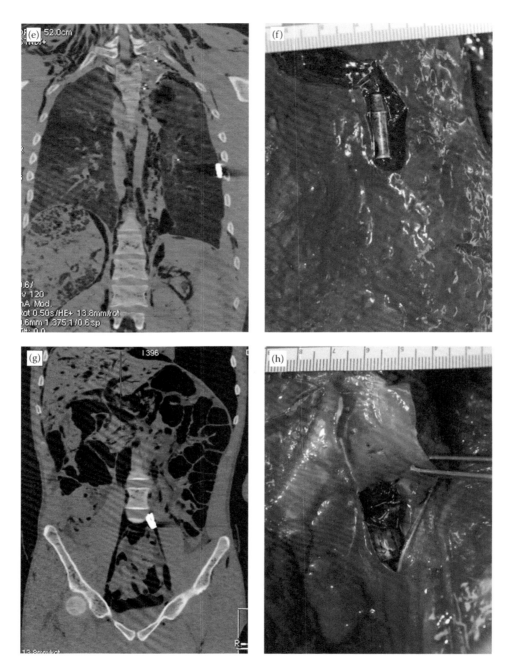

FIGURE 4.12 (*Continued*) Penetrating gunshot of the face with bullet fragmentation shows that bullet fragment location does not always predict the course of the wound track. The bullet entered through the right maxillary sinus and traveled from anterior to posterior, right to left, and downward through the neck. (e) Coronal MDCT of the chest shows a poorly defined descending aorta with adjacent hemorrhage and shows a bullet fragment in the shape of a penetrator in the lateral margin of the left lung. Again noted is a laceration of the left lung with associated hemorrhage and bullet fragments. (f) Autopsy photograph of left lung shows the penetrator located on the pleural surface. (g) Coronal MDCT of the pelvis shows a bullet fragment anterior to L-4 vertebral body. (h) Autopsy photograph shows that the fragment anterior to the L-4 vertebral body is located in the left iliac artery having traveled through the aorta to this location.

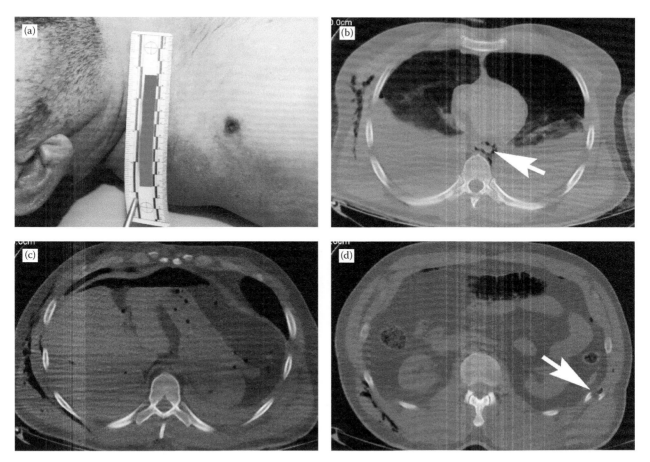

FIGURE 4.13 Penetrating gunshot wound of the torso traveling right to left, anterior to posterior, and downward. (a) Autopsy photograph shows entry wound at right shoulder. (b) Axial MDCT of the chest shows large, bilateral hemothoraxes, left pneumothorax, lung hemorrhages, and posterior mediastinal air collections in the location of the aorta (arrow). These findings suggest major vascular laceration but are not specific. (c, d) Axial images of the abdomen show pneumoperitoneum, extraluminal air collections at the gastric margin, and fractures of the 11th left right rib (arrow in d).

is not positioned perpendicular to the X-ray beam. CT imaging affords the opportunity to study both shape and measurement in three dimensions and overcomes some of the limitations of two-dimensional radiographs. It is possible to ascertain the shape and size of a bullet in multiplanar and three-dimensional MDCT reconstructions with more accuracy. However, metallic artifact on MDCT may degrade edge detail and limit the assessment of shape and dimension (Figure 4.1 and Figure 4.3). Suffice it to say, without presenting the physics of MDCT, the X-ray tube and detector rotation about a metal object creates streak artifact that blurs the margin of metal objects on MDCT images. Consequently, a bullet can exhibit slight variations in shape and measurement on MDCT. Recognizing that both radiography and MDCT have advantages and limitations, we obtain both types of images when evaluating gunshot wound injury in our practice (Figure 4.1 and Figure 4.3).

Bullets are composed of metals that have different composition and atomic number. Different metals may have unique visual and textural or tactile characteristics, but they are often indistinguishable in radiographic attention. Bullets are often encased in a copper or copper alloy jacket. They may also contain a steel penetrator at the tip that is designed to enhance the bullet's ability to penetrate its target. The softer copper deforms more readily than the steel penetrator, and even though they have the same radiographic attenuation, the shape may be a clue to the type of material (Figure 4.3). In cases where metal fragments exhibit a visible difference in attenuation, such as aluminum versus copper or steel, it is possible to see a difference on radiographs.

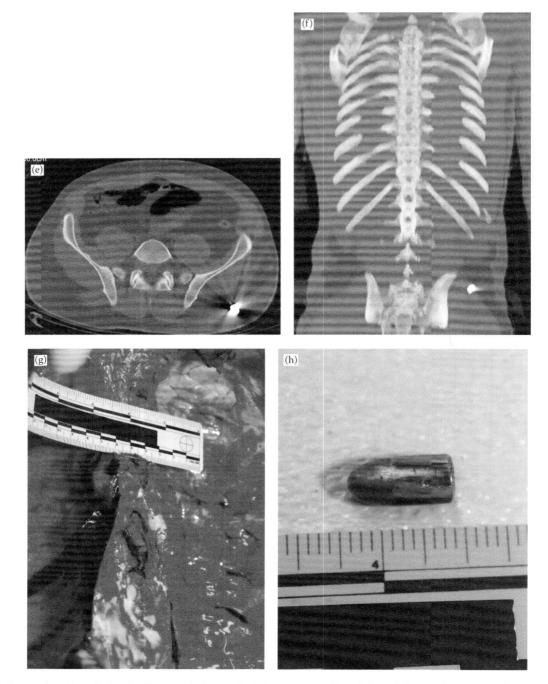

FIGURE 4.13 (*Continued*) Penetrating gunshot wound of the torso traveling right to left, anterior to posterior, and downward. (e, f) Axial and coronal maximum intensity projection images show the rib fracture and projectile posterior to the left iliac bone. (g) Autopsy photograph shows aortic laceration. (h) Photograph of the recovered bullet.

Entry and Exit Wounds

The key to radiographic location of entry and exit wounds is the presence of gas in soft tissue and the disruption of tissue surfaces. Relying on MDCT to locate and classify surface entry and exit wounds can be problematic. Direct visualization of the skin surfaces at autopsy or even by photograph is more accurate. The collapse of the temporary cavities and compression of soft tissue defects produced by the bullet and the position of the body on the CT table limit the detection of entry and exit wounds on MDCT. For example, an exit wound on the back can be compressed and hidden when the body is imaged in the supine position, whereas it may spread open and be identifiable if it is on the anterior chest and not compressed (Figure 4.8).

In visceral organs, entry and exit wounds appear identical. The direction of the bullet path can only be ascertained by analysis of the surrounding bone and soft tissue structures or if metallic bullet fragments or bone fragments have been deposited along the direction of the bullet. The appearance of entry and exit wounds in bone depends on the type of bone and the angle of interaction between the bone and the bullet. Flat bones such as the bones of the skull and pelvis, and bones with high medullary content, such as the vertebral bodies and long bone metaphyses, can retain a "hole-like" fracture defect from a bullet. In contrast, bones with more dense cortical composition, such as the long bone diaphyses, tend to fracture and shatter on impact with a bullet. Consequently, the direction of the bullet may not be as clear when it passes through the diaphyseal regions of long bone (Belkin 1979). If the bone structure is maintained, the surface of the bone where the bullet exited typically shows adjacent bone fragments that have been carried along the direction of travel (Figure 4.14).

Wound Tracks

The radiographic wound track is the visible remnant of the laceration, shock wave, and cavitation created by passage of the projectile. Gas and hemorrhage are the principal features of gunshot wound tracks in soft tissue. Fractures are the manifestations of gunshot wound tracks in bone. There is a high degree of variability in the appearance of gunshot wound tracks depending on the type of tissue the bullet passes through. The gunshot wound track should be determined whenever possible, and a designated direction should be established by identifying the entry wound, bullet pathway, and exit wound (for perforating wounds) (Figure 4.11 and Figure 4.15). Bone findings are the most helpful in determining bullet direction. It is always necessary to account for postmortem changes in organ volume and position which may be secondary to the injury and decomposition because the postmortem location of the gunshot wound track may be shifted. This is most commonly observed in gunshot wounds to the head, where the brain softens and settles dependently from decomposition (Figure 4.5h,i and Figure 4.10a), and in the chest where lung collapse occurs as a result of the gunshot wound or decomposition (Figure 4.16f).

Penetrating Gunshot Wounds

The analysis of penetrating gunshot wounds usually begins with the identification of the retained bullet or metallic fragments. The orientation of the bullet does not always correlate with the direction of entry or the wound track, because bullets may rotate and change direction as they interact with tissue (Figures 4.12 and 4.13). However, the entrance wound may be apparent from disruption of the skin surface, deposition of metal fragments, or subcutaneous gas, and this is helpful. As noted above, entry wounds may not be readily recognized on MDCT, and visual inspection of the body, discussion of the case with the medical examiner, or marking the wound with a radiopaque marker before MDCT may be necessary to establish the location. The next step is to search for a bone abnormality that will indicate a point or points along the bullet track. Positive bone findings in the torso are more likely to be in line with the true track of the bullet than those in the extremities or in highly mobile bones, such as the scapula, which may have been in a different position at the time of injury. Axial MDCT images are useful for detecting sites of abnormality, but coronal, sagittal, and oblique MDCT reconstructions are helpful when matching positive findings to a track (Figure 4.17). To complete the determination of the wound track, the soft tissues should be carefully evaluated to ensure that there is consistency between the entry wound and bone findings. It may be necessary to consider the position of the body at the time of shooting in light of postmortem image findings, which encompass traumatic and decompositional changes (Figure 4.8m and Figures 4.12 and 4.13).

The two most common MDCT features of gunshot wounds are gas collections and hemorrhage. However, gas in blood vessels, organs, and body cavities is also a common feature of decomposition. Decompositional gas is usually symmetric in distribution and follows anatomic tissue planes, as opposed to traumatic gas, which is usually asymmetric, linear, and crosses anatomic boundaries (Levy et al. in press). An additional clue to gunshot wound tracks in soft tissue is the deposition of tiny metal fragments along the course of the track. These metallic fragments are usually too small for evidence recovery but confirm that the bullet has passed through the area (Figure 4.18c).

The ability to identify organ injury varies by organ and degree of decomposition. Careful selection of the optimal CT window and level setting is necessary to observe all of

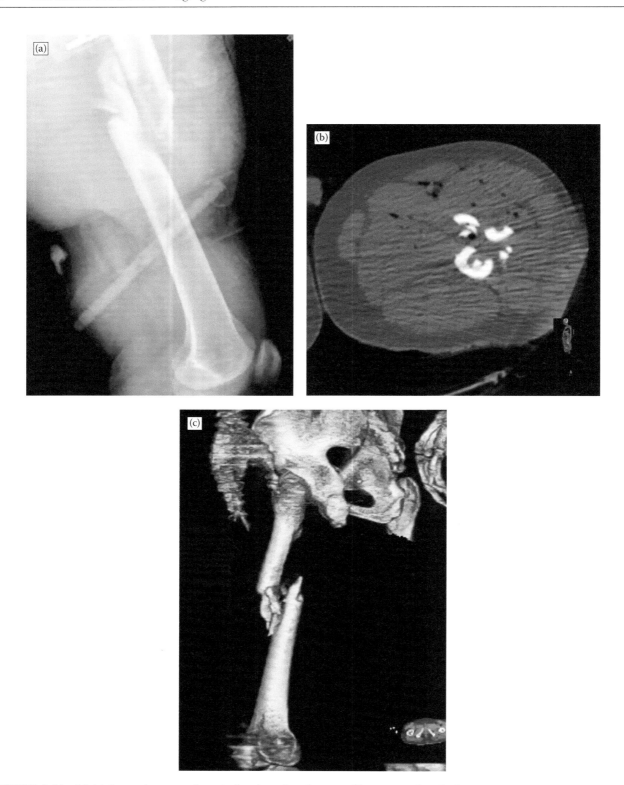

FIGURE 4.14   Multiple gunshot wounds to the leg show that the type of bone injury by a bullet varies by location. (a, b, c) Lateral digital radiograph, axial MDCT, and three-dimensional MDCT of the femoral diaphysis show a gunshot wound that results in a comminuted oblique fracture because the bone is predominantly cortical.

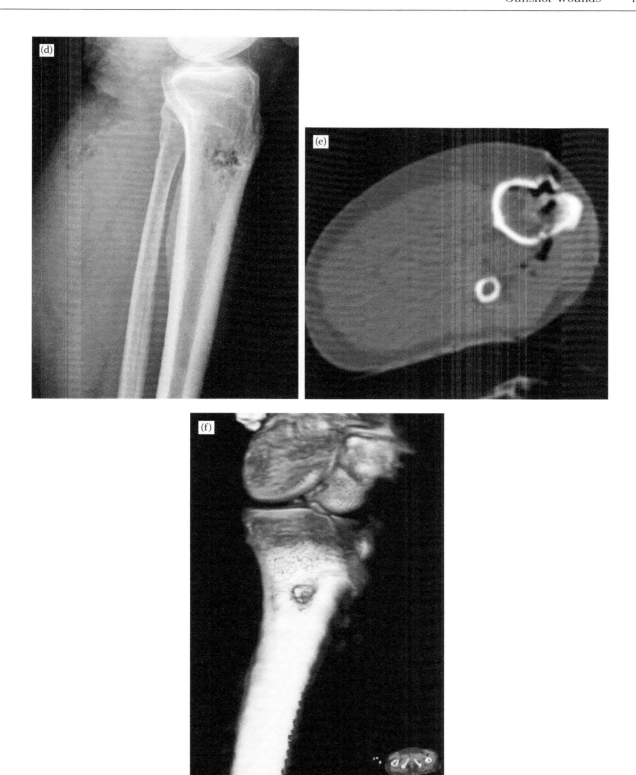

FIGURE 4.14 (*Continued*) Multiple gunshot wounds to the leg show that the type of bone injury by a bullet varies by location. (d, e, f) Lateral digital radiograph, axial MDCT, and three-dimensional MDCT of the tibial metaphysis show a discrete track or hole in the bone because the metaphysis has relatively a thin cortex and more medullary space.

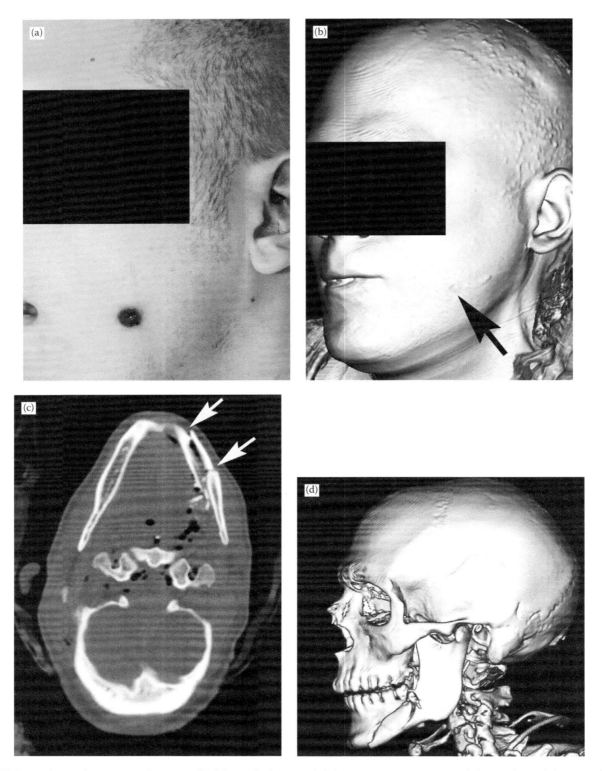

FIGURE 4.15   Perforating gunshot wound of the neck that traveled from anterior to posterior, left to right, and downward. (a) Autopsy photograph and (b) three-dimensional MDCT surface reconstruction show entry wound on the left cheek (arrow). (c, d) Axial and three-dimensional MDCT show fractures of the left mandible.

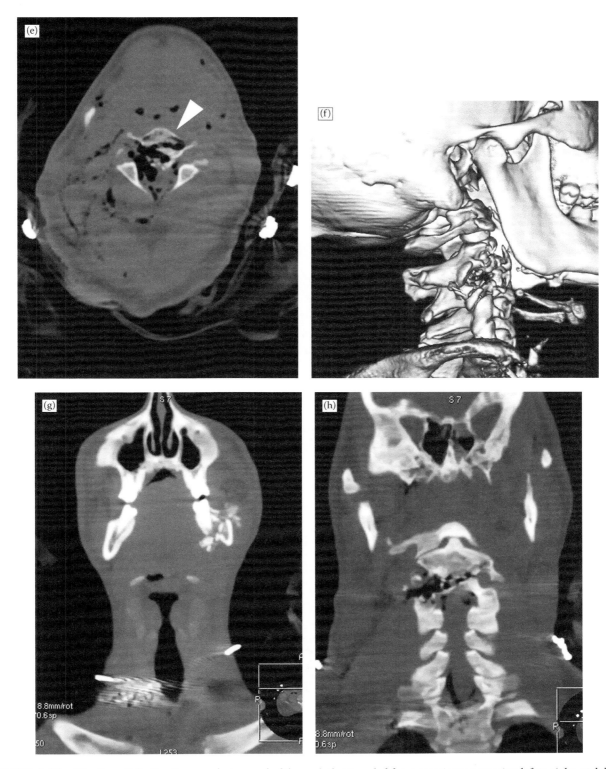

FIGURE 4.15 (*Continued*) Perforating gunshot wound of the neck that traveled from anterior to posterior, left to right, and downward. (e, f) More caudal axial image and three-dimensional reconstruction show cervical fractures (arrowhead). (g, h, i) Coronal MDCT images from anterior to posterior show the left mandible fracture, diagonal fractures through the C2 and C3 vertebral bodies, and a soft tissue wound in the right neck (arrow).

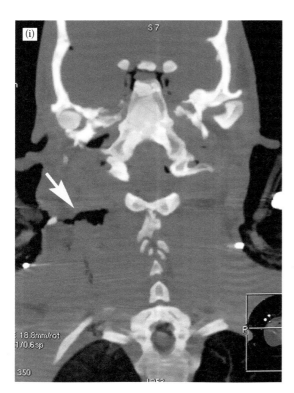

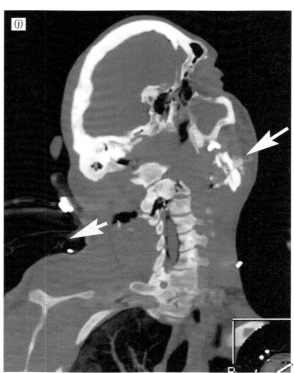

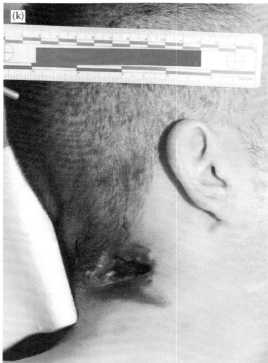

FIGURE 4.15 (*Continued*) Perforating gunshot wound of the neck that traveled from anterior to posterior, left to right, and downward. (g, h, i) Coronal MDCT images from anterior to posterior show the left mandible fracture, diagonal fractures through the C2 and C3 vertebral bodies, and a soft tissue wound in the right neck (arrow). (j) Coronal oblique MDCT shows the wound track (arrows are in the direction of the bullet). (k) Autopsy photograph shows the exit wound.

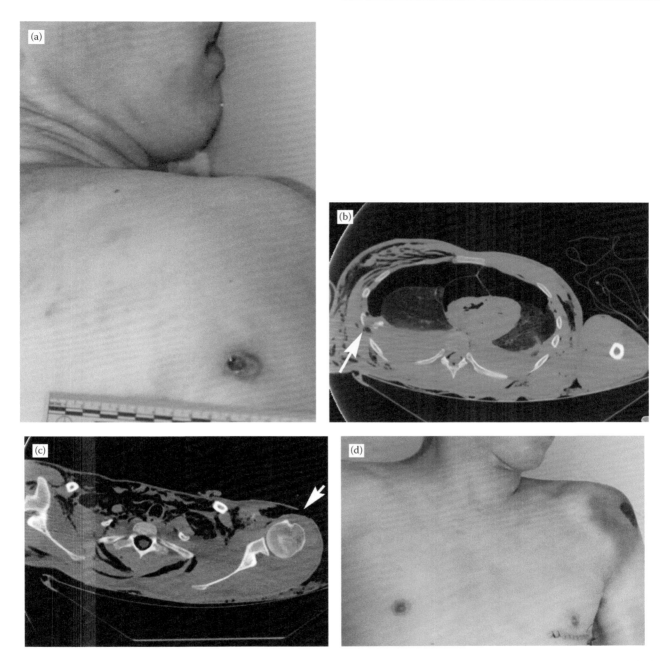

FIGURE 4.16 Perforating gunshot wound of the chest that traveled from posterior to anterior, right to left, and upward. (a) Autopsy photograph shows the entrance wound on the back of the right upper arm. (b) Axial MDCT near the entry wound shows subcutaneous gas, right rib fracture (arrow) with inwardly displaced fracture fragments, and bilateral hemothoraxes. (c) Axial MDCT at the pulmonary apex shows the exit wound on the anterior left shoulder (arrow). (d) Autopsy photograph shows the exit wound on the left shoulder.

these critical MDCT findings. In the brain, decomposition decreases the tissue attenuation and provides good contrast with adjacent hemorrhage, which has high attenuation. Therefore, a gunshot wound track that has significant hemorrhage will be easily recognizable (Figure 4.5h). Similarly, in the lung, hemorrhage along a gunshot wound track is

higher in attenuation compared to the normal surrounding aerated lung. In some cases, a gunshot wound to the chest will cause a significant pneumothorax and accompanying volume loss in the lung such that it becomes difficult to establish the wound track. Hemorrhage may fill alveoli and the pleural space. This can limit volume loss. The resultant

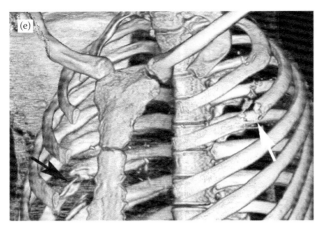
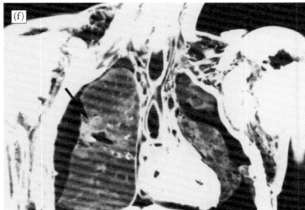

FIGURE 4.16 (*Continued*) Perforating gunshot wound of the chest that traveled from posterior to anterior, right to left, and upward. (e) Three-dimensional MDCT shows fractures of the right seventh (black arrow) and left third (white arrow) ribs which establish the wound track through the thorax. (f) Coronal oblique MDCT viewed in lung windows in the plane of the gunshot wound track shows bilateral pneumothoraxes and part of the wound track through the right lung (arrow) with laceration and surrounding hemorrhage.

hemothorax typically layers in the dependent portions of the affected pleural space (Figure 4.8b,g, and Figure 4.18e).

Organs such as the heart, liver, and spleen are most commonly isoattenuating with hemorrhage on postmortem MDCT. Linear collections of gas within an organ and the disruption of the outer contours of the organ are findings found in gunshot wound tracks through visceral organs. Blood will usually accumulate in the anatomic spaces adjacent to the visceral organs (e.g., pericardium for the heart and retroperitoneal spaces for the kidneys). Once hemorrhage is identified, all surrounding organs should be carefully inspected for more evidence of a wound track. The gastrointestinal tract is the most difficult area to assess for gunshot wound injury because the intestines are often collapsed and small entry and exit wounds may not be apparent. In addition, the presence of pneumatosis is an unreliable sign of injury because it almost always presents as the first stage of putrefactive decomposition. Furthermore, pneumoperitoneum may also be from decomposition. The presence of a pneumoperitoneum is most useful when there are no other decompositional changes present in the body and when there are other findings in the adjacent soft tissues that are compatible with a gunshot wound track (Figure 4.9a and Figure 4.13).

Not all gunshot wound tracks in penetrating injury will be linear, because a bullet or bullet fragment may be diverted when passing through an intermediate target in the body. Bullets may also ricochet from a bone surface

before coming to rest (Harcke et al. 2008). Furthermore, bullets may fragment as they pass through tissue, and secondary wound tracks may be created by the fragments. Although these are typically in the vicinity of the main track, they can become distracting when attempting to define a track.

Finally, there is a tendency to assume that gunshot wounds take place when the victim is upright and that the assailant fires a gun from a horizontal or shallow oblique direction from above or below the victim. But, gunshot wounds can occur from any direction. Steep up and down gunshot wound tracks may pass from head to foot or vice versa if the victim is fired upon from directly above or below in a standing or recumbent position (Figures 4.12 and 4.13).

Perforating Gunshot Wounds

An approach to analysis of perforating gunshot wounds begins with the recognition of all positive findings on the MDCT. It is easiest to recognize metal bullet fragments and bone fractures. The distribution of bone fragments is a clue to the projectile direction of travel (Figure 4.16 and Figure 4.19). Gas collections of significance must be sorted out from decompositional gas. One clue regarding gas collections that we already discussed is asymmetry, but this is not an absolute. Bullets passing through soft tissue do not usually leave a completely traceable gas-filled track. More often there are scattered gas collections at the interfaces between anatomic structures (Figure 4.15). Gas will also dissect through tissue planes

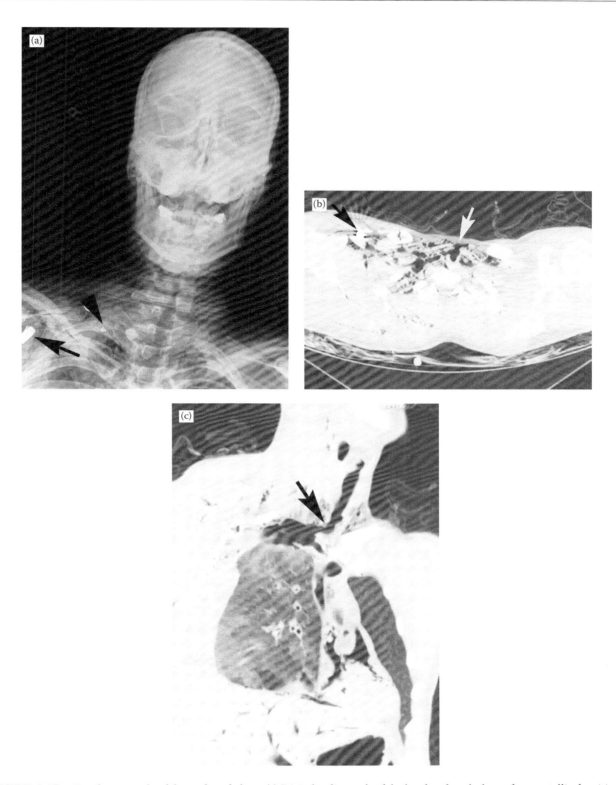

FIGURE 4.17   Gunshot wounds of the neck and chest. (a) Digital radiograph of the head and neck shows four metallic densities to the right of the midline with a fragment below the fractured right clavicle (arrow) and a fragment over the first right rib (arrowhead). There is diffuse soft tissue emphysema. (b) Axial MDCT viewed in lung windows shows an anterior wound (white arrow) and gas in the soft tissues of the upper chest and right clavicular area. A bullet fragment is anterior to the right shoulder (black arrow). (c) Coronal oblique MDCT in plane of wound track viewed in lung windows shows the bullet transected the trachea (arrow). There is a right pneumohemothorax.

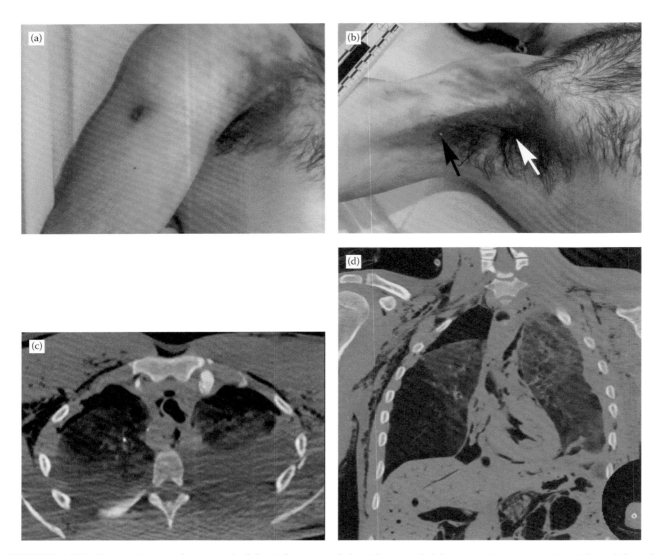

FIGURE 4.18 Penetrating gunshot wound of the right arm and chest that traveled from anterior to posterior, right to left, and upward resulting in a complex gunshot wound track that began in the right arm, crossed the right axilla, passed through the thorax, and fractured the left scapula before coming to rest in the soft tissues posterior to the left shoulder. (a, b) Autopsy photographs show the entry wound in the right upper arm. The bullet exited the arm and entered the chest through the right axilla (shown by arrows in b). At autopsy, flexion and internal rotation of the arm aligned the wound track. (c) Axial MDCT shows gas in the soft tissues of the right chest wall and right intrathoracic chest wall irregularity consistent with a gunshot wound track. The tiny metal fragment in the right lung and the posterior mediastinal air collections indicate the continuation of the track. There is widening of the posterior mediastinum with hemorrhage and the large bilateral hemothoraxes suggesting major vascular damage. (d) Coronal MDCT shows increased attenuation in both upper lungs consistent with hemorrhage, and there are abnormal gas collections in the mediastinum.

and become more dispersed with time. Consequently, gas may not be indicative of the gunshot wound track (Figure 4.16c and Figure 4.17b). Recognition and determination of entrance and exit wounds on MDCT are far more challenging and problematic than at physical autopsy (although it is not a simple task here either). Physical evidence easily observed on external examination is often masked by tissue compression. A gaping gunshot wound on the back closes when the body

is placed supine for MDCT scanning, because the surrounding tissues are pressed together.

Perforating gunshot wounds should have consistency in alignment of the bone and soft tissue abnormalities between the entrance and exit wounds. It may be necessary to imagine the position of the body and anatomic placement of the organs in the body at the time of the shooting to align the soft tissue abnormalities and entrance and

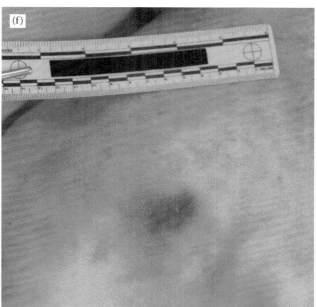

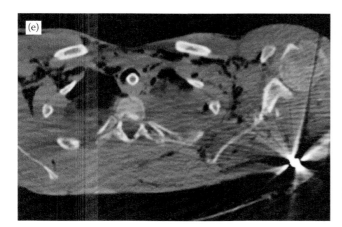

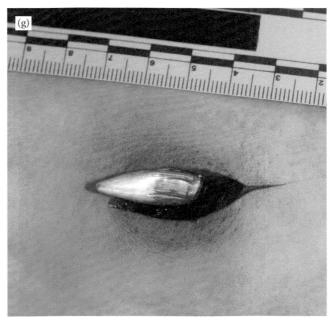

FIGURE 4.18 (*Continued*) Penetrating gunshot wound of the right arm and chest that traveled from anterior to posterior, right to left, and upward resulting in a complex gunshot wound track that began in the right arm, crossed the right axilla, passed through the thorax, and fractured the left scapula before coming to rest in the soft tissues posterior to the left shoulder. (e) Axial MDCT shows the bullet lodged in the subcutaneous soft tissue behind the left scapula, which is fractured. An endotracheal tube is present, and there is bilateral hemopneumothorax. (f) Autopsy photograph shows posterior left shoulder bruising. A hard object was palpable under the skin at autopsy. (g) Photograph of the recovered bullet.

exit wounds (Harcke et al. 2008). Consideration should be given to whether or not the victim was standing, sitting, or lying down and if the victim was in motion or at rest. Other considerations include whether or not the arms and legs were raised or holding something or whether or not the victim was looking straight ahead, left or right, or up or down. Assessment of internal organs may require consideration of the degree of lung inflation, phase of respiration, and shift of organ position with postural changes. Bones are usually the easiest to analyze because they have

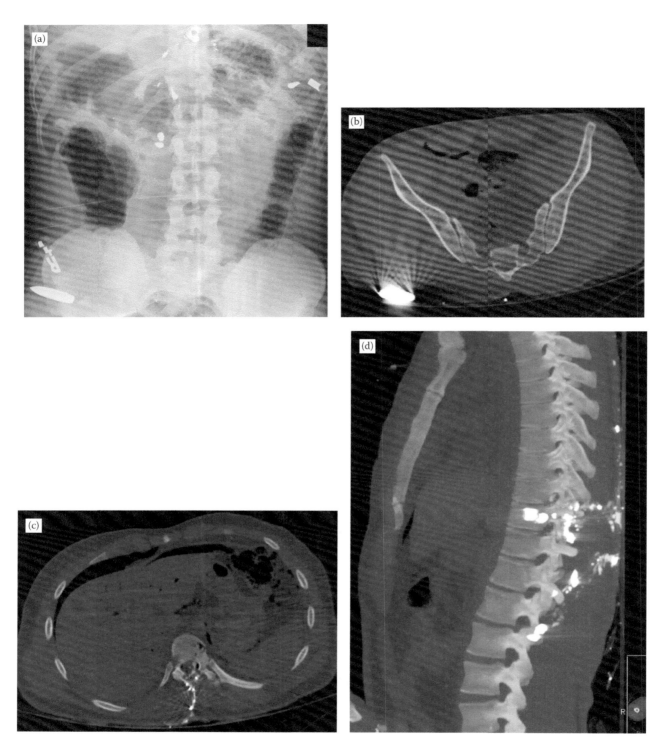

FIGURE 4.19 Multiple gunshot wounds that entered through the back. (a) Digital radiograph of the abdomen and upper pelvis shows multiple metallic fragments in the upper abdomen and an intact bullet over the right iliac bone. (b) Axial MDCT of the pelvis shows the bullet lies outside the body posterior. (c) Axial MDCT at the level of T10 vertebral body shows a bullet track lined by metallic fragments leading through the spinal column. Note the presence of pneumoperitoneum. (d) Sagittal maximum intensity projection shows two wound tracks lined by metallic fragments entering from the back.

the added advantage of suggesting the direction the bullet was traveling. In cases of a single gunshot wound with three skeletal points to align the wound track, it is usually not difficult to establish the exact wound track and verify this at autopsy. Three-dimensional MDCT reconstruction of the skeletal structures can provide assistance in visualizing how a positional change affects alignment (Figure 4.8 and Figure 4.15).

The use of coronal, sagittal, and oblique two-dimensional multiplanar MDCT reconstructions to augment axial scans is usually done simultaneously at an imaging workstation. Findings identified on axial images can and should be confirmed in multiple imaging planes. This is particularly useful for identifying gunshot wound tracks. We found that three-dimensional reconstructions do not add additional diagnostic information but are helpful for understanding and explaining spatial anatomic relationships to nonradiologists. Three-dimensional images may be crucial to the understanding of the case for those who are not accustomed to an axial display of anatomy (Figures 4.5, 4.8, 4.10, 4.14, and 4.16).

Vascular lacerations and disruptions cannot be directly observed on noncontrast MDCT. However, when a wound track passes through the location of a vascular structure and there is evidence of surrounding hemorrhage, it is reasonable to suggest vascular damage. We use the phrase "findings suggestive of (or consistent with) vascular injury" when reporting such cases (Figure 4.20). Postmortem angiography is a technique that can be performed with conventional radiography or MDCT to document vascular injury (Figure 4.7 and Figure 4.21). We incorporate postmortem angiography with the autopsy to facilitate vascular access leading directly to or from the area of interest. Postmortem angiography may be performed with an antegrade or retrograde injection of diluted radiographic contrast material. Other published postmortem angiographic methods use different radiographic contrast agents and infusion techniques (Grabherr et al. 2007).

Special Cases

There are a number of variations of basic penetrating and perforating gunshot wounds that produce special characteristics. Gunshot wounds of the skull and facial bones typically produce extensive fractures that extend well beyond the points of entry and exit. High-velocity bullets passing into or through the skull result in complex, comminuted fractures that often interconnect because gunshot wounds to the skull create high intracranial pressures (Brogdon 1998). Smith et al. assert that radiating linear skull fractures and concentric heaving skull fractures can be used to determine the direction of a gunshot track. Gunshot entrance wounds may create long linear skull fractures that are not arrested by preexisting fractures or concentric heaving skull fractures that are more fragmented and have elongated radial fractures emanating from them. In contrast, fractures associated with exit wounds have radial and heaving fractures of lesser magnitude that may be arrested by preexisting fractures (Smith et al. 1987).

The inner and outer tables of the skull may show different characteristics when a bullet strikes tangentially as opposed to perpendicular bullet entrance. The perpendicular entrance fracture should show a punched-out hole on the outer table and internal beveling on the inner table. Similarly, at the exit fracture, the inner table is punched out and the beveling occurs on the outer table (Spitz et al. 2006). Tangential or keyhole and grazing gunshot wounds that interact with the curvature of the skull can have entry and exit wounds close together and overlapping (Figure 4.22). In the overlapping keyhole fracture, one end will show characteristics of an entrance defect, and the other end will show a bevel pattern consistent with an exit defect (Coe 1982, Harcke et al. 2007).

Perhaps the most challenging gunshot wound cases are those with multiple gunshot wounds. The possibilities include varying number and complexity as well as mixtures of penetrating and perforating wounds. Wound tracks may cross or commingle, and they may originate from more than one weapon. The wound entry can be from a different direction for each shot if the victim's body is in motion during the shooting. Other, more complicated possibilities include that the shots may have been fired at different times or that there may be a mixture of both antemortem and postmortem gunshot wounds. Radiographic and MDCT analysis of these complex cases is best done in conjunction with the medical examiner's external examination. Knowledge of number and nature of the wounds provides the radiologist a framework to

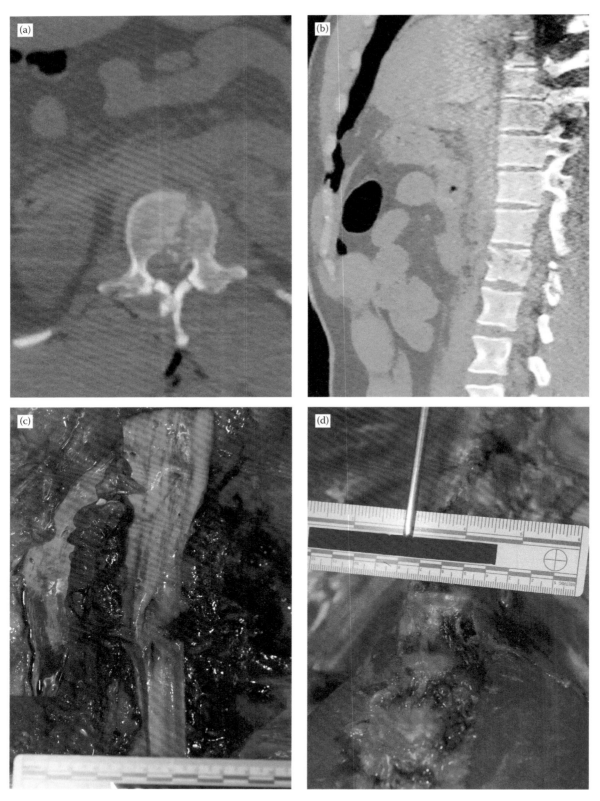

FIGURE 4.20   Perforating gunshot wound of the chest and abdomen that traveled from anterior to posterior, left to right, and downward. (a, b) Axial and sagittal MDCT show a fracture of the L1 vertebral body and retroperitoneal hemorrhage, which suggests major vascular injury. (c) Autopsy photograph shows aortic laceration and surrounding hemorrhage. (d) Autopsy photograph shows fracture and hemorrhage of the L1 vertebral body.

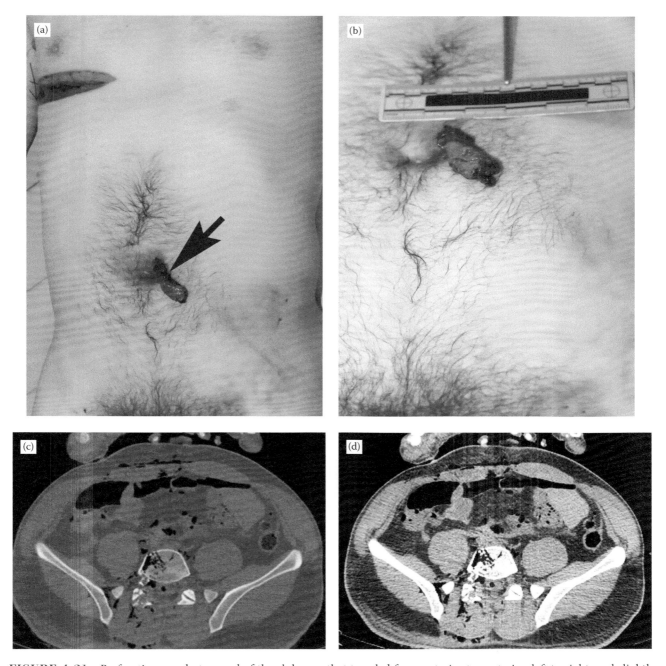

FIGURE 4.21   Perforating gunshot wound of the abdomen that traveled from anterior to posterior, left to right, and slightly downward. (a, b): Autopsy photographs of the entry wound show bowel herniation (arrow in a). A thoracotomy incision made during the resuscitation attempt is also present in (a). (c, d, e) Axial and sagittal oblique (in the plane of the wound track) MDCT show the anterior entry wound and a fracture defect in L5. Note the bone fragments are posterior.

use when studying the images. When the number of entry wounds exceeds the number of exit wounds, the implication is that both penetrating and perforating wounds are present and bullets should be present. However, it is also possible in rare cases that more than one bullet exited through the same wound. If the number of exit wounds

exceeds the number of entry wounds, it may be due to fragmentation of a bullet or more than one bullet entering through the same entrance wound. The medical examiner will be much more accurate in estimating number of shots fired and associating wounds from the same bullet. Grazing wounds and wounds passing through an arm or

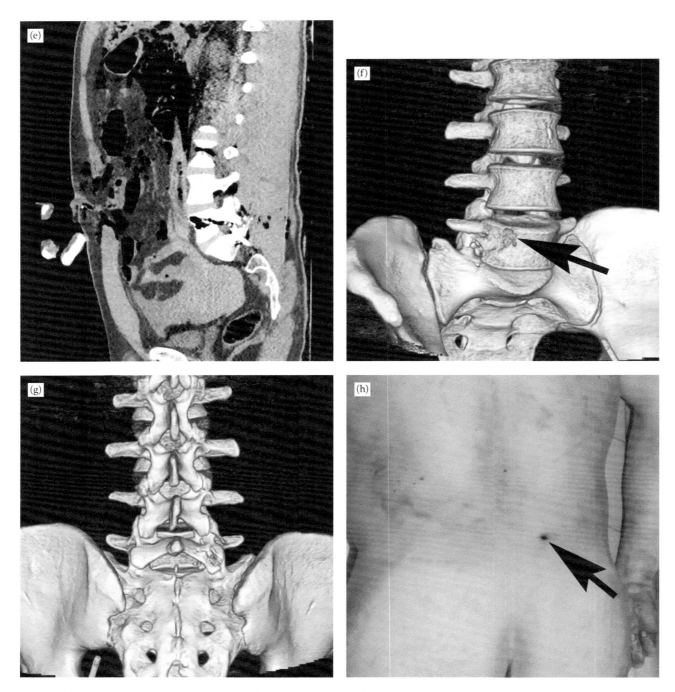

FIGURE 4.21   (*Continued*) Perforating gunshot wound of the abdomen that traveled from anterior to posterior, left to right, and slightly downward. (c, d, e) Axial and sagittal oblique (in the plane of the wound track) MDCT show the anterior entry wound and a fracture defect in L5. Note the bone fragments are posterior. (f, g) Three-dimensional MDCT shows the entry defect at the L5 body (arrow) and the exit at the left inferior L5 facet. (h) Autopsy photograph shows exit wound on right lower back (arrow).

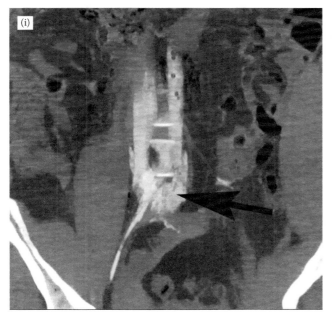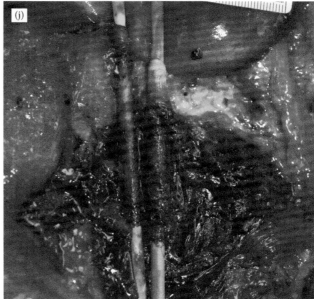

FIGURE 4.21 (*Continued*) Perforating gunshot wound of the abdomen that traveled from anterior to posterior, left to right, and slightly downward. (i) Coronal MDCT from postmortem angiogram done by retrograde injection of the right iliac artery. There is intravenous contrast extravasation at the aortic bifurcation (arrow) and filling of the inferior vena cava. (j) Autopsy photograph shows laceration of the aorta (pink rod) and vena cava (blue rod) with surrounding hemorrhage.

leg and reentering the torso (or vice versa) are confounding for the radiologist (Harcke et al. 2007, Levy et al. 2006). Both the radiologist and medical examiner will have difficulty when more than one shot has the same point of entry or exit. As the medical examiner attempts to match entry and exit for perforating wounds, MDCT data can be very helpful. MDCT is also very helpful because anatomic regions that are not routinely seen during initial autopsy dissection, such as the paraspinal regions, can easily be evaluated, and observation of abnormalities in these areas will often help confirm or reject suspected gunshot wound paths (Figure 4.20).

CONCLUSIONS

The application of MDCT to the postmortem examination of gunshot wounds provides the medical examiner with information that saves time and increases the accuracy of the autopsy. It is still recommended that conventional radiographs be obtained in order to characterize bullet and fragment detail. Using MDCT images to guide the recovery of bullets and ballistic fragments at autopsy enhances the efficiency of the autopsy process and obviates the need for a C-arm to localize bullets. Identification of entry and exit wounds is best done in conjunction

with the external examination of the medical examiner. Multiplanar two- and three-dimensional MDCT images are used to determine gunshot wound tracks that can be correlated with anatomic findings at autopsy. Skeletal pathology is exquisitely detailed on MDCT, which permits the visualization of structures that are not routinely examined at autopsy because of inaccessibility. There are limitations in the evaluation of vascular and solid organ injury; however, angiography can be used to define damage and limit dissection. An additional benefit of MDCT is that the scans can be used as evidence in court and in most cases will be more readily admitted into evidence over a bloody or otherwise inflammatory and possibly prejudicial photograph.

REFERENCES

Belkin, M. 1979. Wound ballistics. *Prog Surg* 16: 7–24.

Brogdon, B. G. 1998. *Forensic radiology,* Boca Raton, FL: CRC Press.

Centers for Disease Control and Prevention. 2008. WISQARS injury mortality reports, 1999–2005. http://webappa.cdc.gov/sasweb/ncipc/mortrate10_sy.html (accessed August 13, 2008).

Coe, J. I. 1982. External beveling of entrance wounds by handguns. *Am J Forensic Med Pathol* 3: 215–219.

Denton, J. S., Segovia, A., and Filkins, J. A. 2006. Practical pathology of gunshot wounds. *Arch Pathol Lab Med* 130: 1283–1289.

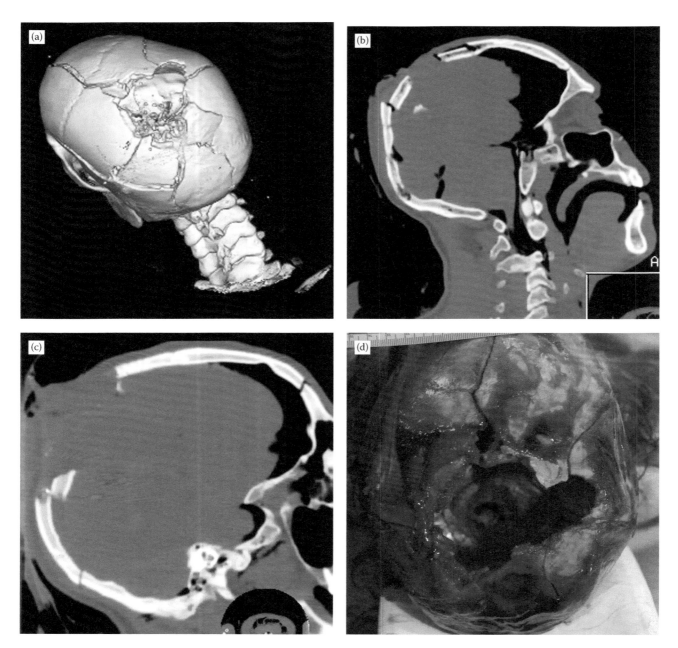

FIGURE 4.22   Tangential perforating gunshot wound of the skull creating a classic keyhole fracture. The bullet traveled from anterior to posterior, right to left, and downward. (a) Three-dimensional MDCT of the skull shows a keyhole configuration of the entry and exit fractures in the skull. (b) Sagittal MDCT at the level of the entry wound shows internal beveling. (c) Sagittal reconstruction at exit shows external bevel. (d) Autopsy photograph shows the keyhole defect and multiple skull fractures.

Di Maio, V. J. M. 1999. *Gunshot wounds: practical aspects of firearms, ballistics, and forensic techniques,* Boca Raton, FL: CRC Press.

Grabherr, S., Djonov, V., Yen, K., Thali, M. J., and Dirnhofer, R. 2007. Postmortem angiography: review of former and current methods. *AJR Am J Roentgenol* 188: 832–838.

Harcke, H. T., Levy, A. D., Abbott, R. M. et al. 2007. Autopsy radiography: digital radiographs (DR) vs multidetector computed tomography (MDCT) in high-velocity gunshot-wound victims. *Am J Forensic Med Pathol* 28: 13–19.

Harcke, H. T., Levy, A. D., Getz, J. M., and Robinson, S. R. 2008. MDCT analysis of projectile injury in forensic investigation. *AJR Am J Roentgenol* 190: W106–W111.

Levy, A. D., Abbott, R. M., Mallak, C. T. et al. 2006. Virtual autopsy: preliminary experience in high-velocity gunshot wound victims. *Radiology* 240: 522–528.

Levy, A. D., Harcke, H. T., and Mallak, C. T. 2010. Postmortem imaging: MDCT features of postmortem change and decomposition. *Am J Forensic Med Pathol* 31: 12–17.

Peterson, G. F., and Clark, S. C. 2006. Forensic autopsy performance standards. National Association of Medical Examiners Annual Meeting (October 16, 2006), San Antonio, TX.

Smith, O. C., Berryman, H. E., and Lahren, C. H. 1987. Cranial fracture patterns and estimate of direction from low velocity gunshot wounds. *J Forensic Sci* 32: 1416–1421.

Spitz, W. U., Spitz, D. J., and Fisher, R. S. 2006. *Spitz and Fisher's medicolegal investigation of death: guidelines for the application of pathology to crime investigation,* Springfield, IL: Charles C Thomas.

Thali, M. J., Yen, K., Vock, P. et al. 2003. Image-guided virtual autopsy findings of gunshot victims performed with multi-slice computed tomography and magnetic resonance imaging and subsequent correlation between radiology and autopsy findings. *Forensic Sci Int* 138: 8–16.

Wilson, A. J. 1999. Gunshot injuries: what does a radiologist need to know? *Radiographics* 19: 1358–1368.

Chapter 5
Blunt Force Injury

FORENSIC PRINCIPLES

Blunt force injury occurs when a blunt object strikes the body or the body impacts against a blunt object or surface. It is the most common form of lethal and nonlethal trauma. Motor vehicle accidents, motor vehicle and pedestrian accidents, falls, aviation crashes, and homicide are common causes of blunt force injury. Because the force and mechanism of blunt force injury are highly variable, there is a broad spectrum of blunt force wounds. Blunt force wounds are classified into four major categories: abrasions, contusions, lacerations, and skeletal fractures. The severity of these injuries depends on the amount of force inflicted on the body, the amount of surface area over which the force is delivered, the portion of the body affected, the type of weapon or object and its impact on the body, and the time over which the force is delivered (Davis 1998).

Wound Classification

Abrasions are wounds produced by scraping or removing the epidermis when there is friction or movement of the skin against a rough surface. Abrasions may be superficial or deep and may be subclassified as scrapes, impact abrasions, or patterned abrasions. Impact abrasions occur when force is directed perpendicular to the body. They are usually focal and located over bony prominences. Patterned abrasions are characterized by a specific pattern from a material or object that marks the skin. At autopsy, it is important to differentiate wounds that were inflicted on the victim antemortem, perimortem, or postmortem. Antemortem abrasions are typically reddish brown. In contrast, postmortem abrasions lack coloration because of the absence of vascular perfusion (Di Maio and Di Maio 2001). Those that occur near the time of death may have subtle changes.

Contusions, or bruises, are hemorrhages in the skin, soft tissue, or internal organs. They are initially dark red, reddish blue, or purple and change color over time, progressing to brownish yellow and then light yellow as the hemoglobin within the contusion breaks down. Postmortem contusions occur when a large force is delivered to the body within a few hours after death. These can be differentiated from antemortem contusions by the lack of microscopic evidence of vital reaction (Davis 1998). Timing of infliction of a contusion is very difficult to determine due to the variable timing of wound healing.

Lacerations are tears in tissue. Similar to contusions, they may occur in the skin, soft tissue, or internal organs. Lacerations occur when tissues are stretched from a shearing or crushing force (Di Maio and Di Maio 2001). They commonly have irregular, contused, and abraded margins. Bridging tissue is the incomplete tearing of the connective tissue and vascular structures within a laceration. The finding of bridging tissue within a wound is helpful to distinguish a laceration from a knife wound or a wound from another sharp object, which typically contains bridging tissue because knives and sharp objects cut rather than tear tissue. It is not unusual to find foreign material embedded in lacerations and deep abrasions when the injury is from severe trauma, such as a motor vehicle accident. Amputations are severe forms of lacerations. Skeletal fractures are lacerations of bone.

AUTOPSY FINDINGS

The autopsy and forensic investigation of death from blunt force injury requires an accurate account of the fatal incident, data from the scene investigation, and careful analysis and dissection of all wounds. Blunt force injuries are commonly divided into two categories: impact injuries and acceleration-deceleration injuries. Both types of injuries may coexist, especially in motor vehicle and aviation accidents. Head injuries are the most common cause for fatal injury in motor vehicle accidents, followed by thoracic injuries (Swierzewski et al. 1994, Wiegmann and Taneja 2003). Other injuries such as burns, drowning, asphyxiation, and blast may be superimposed on blunt force injury, making the cause of death difficult to determine in some cases. In homicides involving blunt trauma, wound and injury pattern analysis may help to estimate or determine the type of weapon used to inflict trauma. Autopsy may reveal additional wounds such as defensive wounds that may be helpful in determining the circumstances of death.

There are several considerations that should be kept in mind when examining a victim of blunt trauma. The presence or absence of external signs of blunt trauma does not indicate the extent or severity of internal injury (Davis 1998). It is possible to have significant internal injury with no evidence of abrasions, bruising, contusions, or lacerations on external examination. Consideration should also be given to preexisting or newly discovered natural diseases that may have contributed to death when there has been mild or moderate trauma that may not have caused death in the absence of underlying natural disease.

Craniocerebral Injuries

Blunt force trauma to the head may occur from the impact of an object striking the head or from sudden movement of the head during the trauma. The impact of an object on the head may produce soft tissue injury, fractures, contusions, or intracranial hemorrhage (Figures 5.1 and 5.2). Sudden movement of the head results in acceleration or deceleration injuries, namely hemorrhage in the form of subdural hematomas or diffuse axonal injury (Di Maio and Di Maio 2001).

Linear fractures are the simplest form of skull fracture (Figure 5.2). With increasing force or velocity of impact, skull fractures become more complex: circular, stellate, comminuted, or depressed (Figure 5.3). Basilar skull fractures are commonly present in fatal head, neck, and facial trauma. They are associated with hemorrhage along the petrous ridges and clivus. Cortical contusions begin as areas of petechial hemorrhage that may coalesce into larger collections. Contusions lead to necrosis within the brain parenchyma. At autopsy, contusions appear as focal areas of hemorrhage or necrosis. They may occur with or without an associated skull fracture. Contusions are usually located at or near the crest of a gyrus and are most commonly found in the frontal and temporal lobes (Graham et al. 1995). Cortical contusions may also occur in a coup-contracoup pattern (Figure 5.2). The coup contusion is at the site of the impact, and the associated contracoup contusion is directly opposite the point of impact. A line drawn from the center of the coup lesion to the center of the contracoup lesions corresponds to the direction of impact (Morrison et al. 1998). Contusions on the superior margin of the cerebral hemispheres in the parasagittal white matter, orbital surfaces of the inferior frontal lobe,

and basal surface of the temporal lobe are often referred to as *gliding contusions* (Morrison et al. 1998). They occur from rotational acceleration of the head, which is common in motor vehicle accidents. Cerebral lacerations and intraparenchymal hematomas may also occur at the site of impact in very severe trauma.

Subdural hematomas and diffuse axonal injury are classic acceleration-deceleration injuries. Subdural hematomas arise from tearing of the bridging veins that cross the subdural space. The blood that collects in the subdural space may also contain cerebrospinal fluid if the arachnoid membrane tears with the bridging veins. At autopsy, subdural hematomas may be seen in the acute, subacute, or chronic stages. Often there is an associated brain injury, which is ultimately the significant contributor to the mechanism of death in many cases (Figures 5.4 and 5.5).

Diffuse axonal injury (also called *shearing injury*) is caused by rotational forces on the brain or acceleration-deceleration injury. At autopsy, the brain may appear grossly normal or show signs of edema (Figure 5.4). Diffuse axonal injury is identified microscopically when there is evidence of diffuse axonal swelling. The cerebral hemispheres, corpus callosum, brain stem, and less commonly, the cerebellum, are typical locations of injury. In severe cases, there is disruption of the axons with grossly visible hemorrhage (Adams et al. 1989).

Focal or diffuse subarachnoid hemorrhage is present in most cases of moderate to severe head trauma (Figure 5.6). The hemorrhage is the result of tearing of blood vessels on the surface of the central nervous system. Subarachnoid hemorrhage may originate at the point of injury or diffuse over the surface of the brain, brain stem, and spinal cord due to gravity and the position of the body after death. The diagnosis of subarachnoid hemorrhage at autopsy is often difficult because blood may enter the subarachnoid space by the removal of the top of the calvarium or may accumulate from decomposition (Di Maio and Di Maio 2001).

Thoracoabdominal Injuries

Thoracic injuries account for 25% of trauma-related deaths in the United States and are an important component of multisystem blunt trauma (Kaewlai et al. 2008). A spectrum of injuries in the soft tissues and bones of the chest wall, heart and major blood vessels, lungs, and diaphragm

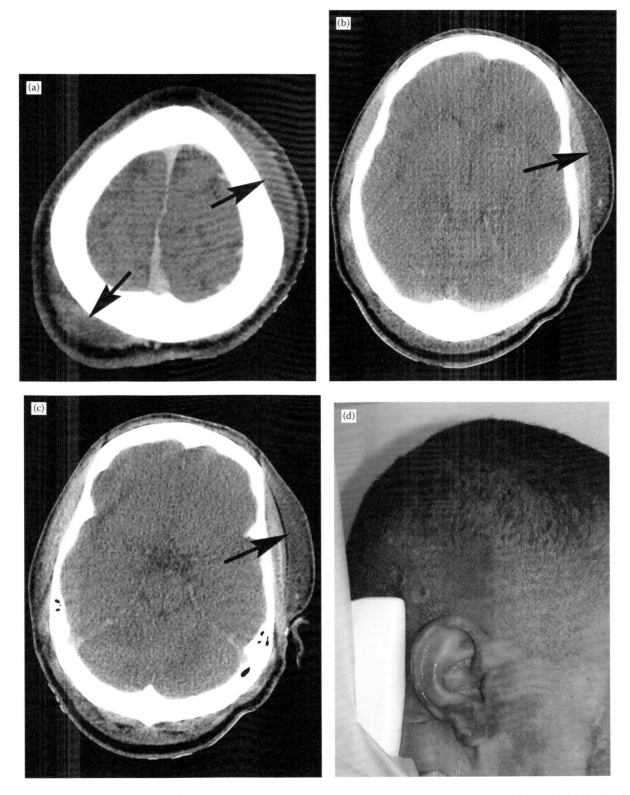

FIGURE 5.1 A 22-year-old man died from multisystem blunt trauma in a motor vehicle accident. (a, b, c) Axial MDCT of the brain shows left parietal-temporal and right occipital region subgaleal hematomas (arrows). No fractures or brain abnormalities were identified. The brain and skull were normal at autopsy. (d) Autopsy photograph shows abrasions and contusions of the posterior and temporal regions of the scalp and abrasions of the face and ear.

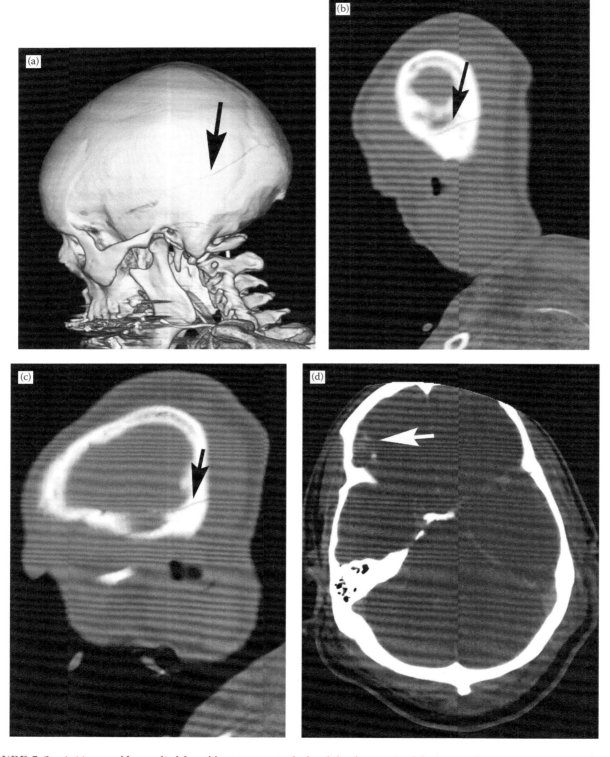

FIGURE 5.2   A 44-year-old man died from blunt trauma to the head that he sustained during an altercation. He was struck in the face and fell to the floor, hitting his head on the floor. (a, b, c) Three-dimensional and sagittal MDCT show a nondisplaced linear skull fracture (arrows) in the left temporal-occipital skull. (d) Axial MDCT of the brain shown in a subdural window shows a right frontal contracoup contusion (arrow) and diffuse subarachnoid hemorrhage.

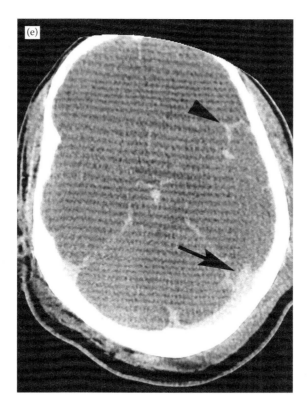

FIGURE 5.2 (*Continued*) A 44-year-old man died from blunt trauma to the head that he sustained during an altercation. He was struck in the face and fell to the floor, hitting his head on the floor. (e) Axial MDCT of the brain shown in a soft tissue window shows diffuse subarachnoid hemorrhage (arrowhead) and a coup contusion in the posterior left temporal lobe (arrow). At autopsy, there was a small left temporal epidural hematoma that is not identified on the MDCT.

may occur. During autopsy, it is important to distinguish cardiopulmonary resuscitation (CPR) injuries from blunt traumatic injuries when evaluating the chest wall. CPR-associated rib fractures are most commonly identified on the anterior portion of the ribs (anterior buckle fractures) and typically do not have associated hemorrhage. Sternal fractures from resuscitation are most common in the lower portion of the sternum (Lederer et al. 2004). In contrast, traumatic rib fractures usually have associated hemorrhage and may cause contusion or laceration of the underlying pleura or lung (Figure 5.7). Blunt traumatic fractures of the sternum, particularly in motor vehicle accidents, are typically transverse fractures at the third intercostal space. Pneumothorax, hemothorax, and pulmonary laceration and contusion may occur as a consequence of rib fracture. Other associated injuries include airway injuries when there is upper chest trauma, and liver, spleen, and diaphragm injury when the lower ribs are fractured.

Blunt force injury to the chest and upper abdomen are common in motor vehicle accidents because compressive forces impact the chest when the driver hits the steering wheel or when passengers impact the dashboard, windshield, or seats. Seatbelts also cause injuries with a pattern that corresponds to the position of the belt. The sudden deceleration of the body at impact with a static structure causes the heart and great vessels to be forcefully pulled away from the posterior chest where the aorta is anchored by the ligamentum arteriosum. Consequently, laceration of the aorta just distal to the origin of the left subclavian artery at the ligamentum arteriosum is the most common site for aortic laceration. Aortic laceration accounts for 10% to 15% of motor vehicle accident deaths in the United States (Kaewlai et al. 2008). At autopsy, there is periaortic and mediastinal hemorrhage and focal laceration on the intimal surface of the aorta or complete transection of the aorta.

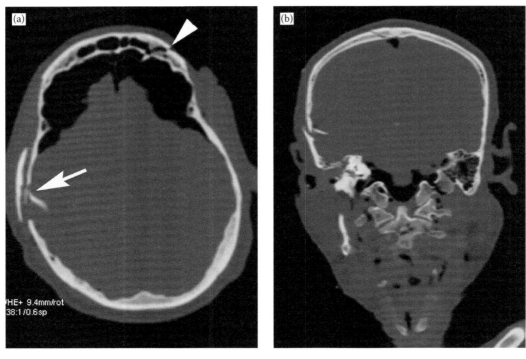

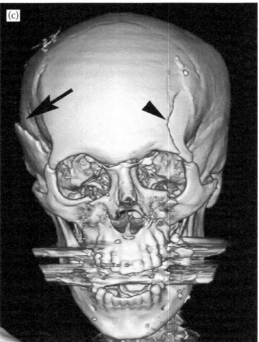

FIGURE 5.3 A 61-year-old man died from head trauma in a motor vehicle accident. (a, b) Axial and coronal MDCT shows a depressed skull fracture of the right parietal temporal region (arrow) and fracture of the left frontal sinus (arrowhead) with overlying soft tissue defect. The brain has retracted from decomposition, and there is decompositional gas within the cranium. (c) Three-dimensional MDCT shows the right-sided depressed skull fracture (arrow) and a left frontal fracture that extends to the orbit (arrowhead). There is streak artifact from dental restoration.

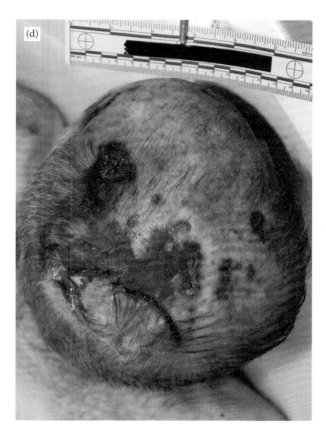
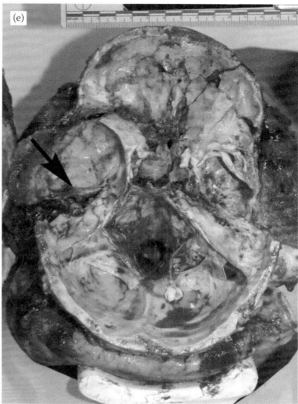

FIGURE 5.3 (*Continued*) A 61-year-old man died from head trauma in a motor vehicle accident. (d) Autopsy photograph of the head shows multiple abrasions and lacerations with exposure of the skull in the right parietal vertex region. (e) Autopsy photograph of the opened skull shows comminuted fractures of the anterior cranial fossa, left middle cranial fossa (arrow), and a linear fracture extending through the right middle cranial fossa to the right temporal region. The autopsy photograph orientation for (d) and (e) is reversed to match the orientation of the computed tomography.

In motor vehicle accidents, cardiac injuries are less common than aortic. Myocardial contusions and lacerations as well as lacerations of the pericardium may occur. Contusions may simulate myocardial infarction clinically, causing arrhythmias or minor changes in the electrocardiogram. Grossly, contusions are focal hemorrhagic areas in the myocardium. Lacerations may involve any portion of the heart and the pericardium. They tend to be linear in configuration and may be superficial or extend through the myocardium. Depending upon the severity of the laceration, there may be an accompanying mediastinal hematoma, hemothorax, or hemopericardium (Figure 5.7). Cardiac tamponade may cause death with as little as 150 cc of blood in the pericardium (Di Maio and Di Maio 2001).

Traumatic rupture of the diaphragm is usually associated with multiorgan injury. It occurs when blunt force is applied to the lower chest or upper abdomen. The left hemidiaphragm is more frequently ruptured than the right. Abdominal viscera usually protrude through the laceration into the thorax (Figure 5.8).

Injury to abdominal organs is also common. Although abdominal organ injury is generally not the cause of death, it may contribute to an individual's morbidity and mortality in multisystem trauma. Hemorrhage from laceration or rupture of the liver, spleen, kidneys, or major vascular structures in the abdomen may be the cause of death when there is significant force applied to the abdomen resulting in massive intraperitoneal or retroperitoneal hemorrhage. Anatomically, the spleen is protected in the posterior left upper abdomen, whereas the liver is partially protected by the ribs. Laceration or rupture of the spleen is frequently associated with rib fractures and significant force to the left upper abdomen. In contrast, liver lacerations may occur with or without rib fracture. Spleen and liver lacerations may involve tearing of the capsule and parenchyma or may be intraparenchymal with an intact

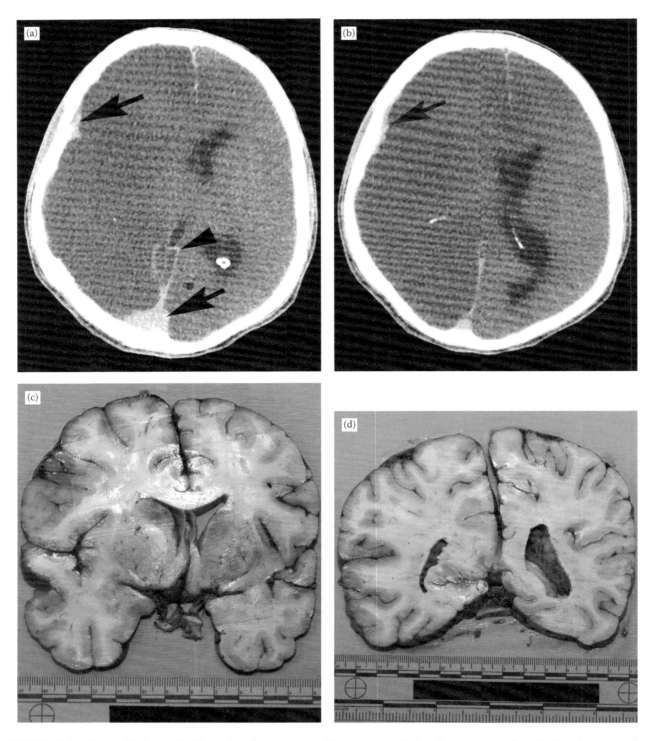

FIGURE 5.4 Unidentified man died from closed head trauma. He was reportedly involved in a struggle and fell to the ground, hitting his head. (a, b) Axial MDCT of the brain shows extra-axial hemorrhage in the right frontotemporal and occipital regions (arrows). At autopsy, both hemorrhages were subdural in location. MDCT also shows subarachnoid hemorrhage (arrowhead) and swelling of the brain, which is more severe in the right hemisphere than the left. There is midline shift toward the left, subfalcine herniation, and enlargement of the left lateral ventricle. (c, d) Autopsy photographs of a coronal section of the brain show diffuse cerebral edema that is more severe on the right. There is leftward midline shift, compression of the right lateral ventricle, and enlargement of the left lateral ventricle. Histology revealed diffuse axonal injury. The autopsy photograph orientation is reversed to match the orientation of the computed tomography.

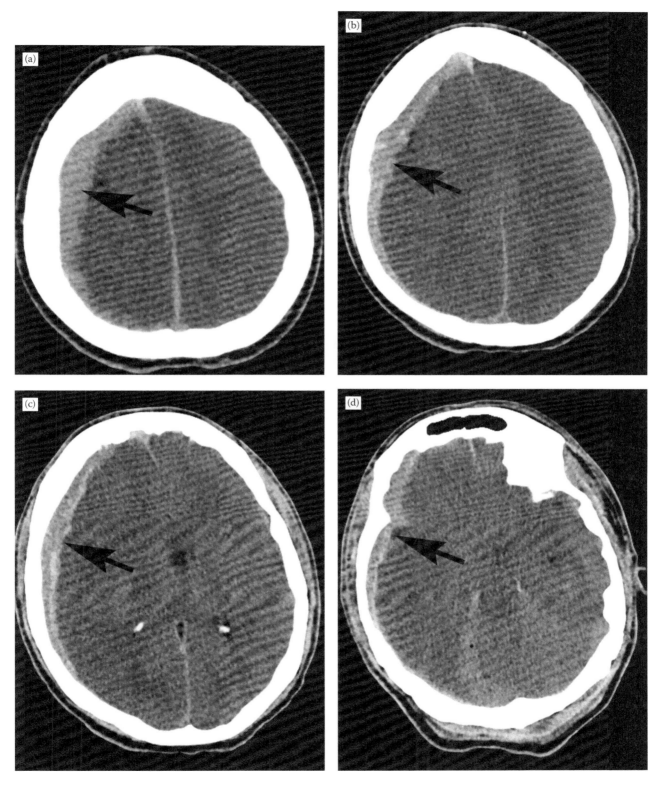

FIGURE 5.5 Motor vehicle accident victim who died from multisystem blunt trauma. (a, b, c, d) Axial MDCT of the brain shows a hyperattenuating acute right subdural hematoma (arrow). The subdural hematoma is heterogeneous in some areas that likely reflect mild decompositional change. There is diffuse edema of the right cerebral hemisphere, compression of the ventricular system, and subfalcine herniation.

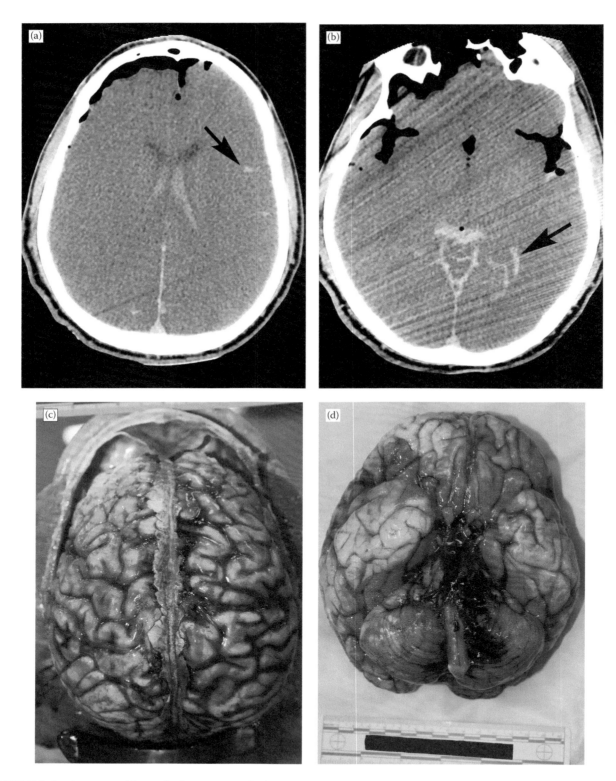

FIGURE 5.6   A 22-year-old man died in a motor vehicle accident from multisystem blunt trauma. (a, b) Axial MDCT images of the brain show diffuse subarachnoid hemorrhage (arrow in a) and focal subarachnoid hemorrhage adjacent to the left cerebellum. A small amount of intraventricular hemorrhage is also present (arrow in b). The intracranial gas and loss of gray and white matter differentiation is due to decomposition. (c, d) Autopsy photographs of the brain show diffuse subarachnoid hemorrhage and a focal subarachnoid collection of blood adjacent to the left cerebellum. The orientation of the autopsy photograph (c) is reversed to match the orientation of the computed tomography.

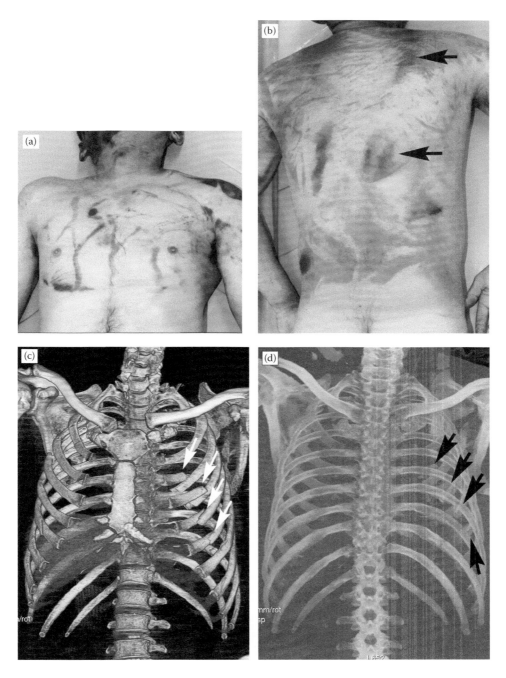

FIGURE 5.7   A 39-year-old male pedestrian was struck by a motor vehicle and died of fatal head and chest injuries. (a, b) Autopsy photographs of the back and chest show patterned abrasions of a fabric imprint on the midsection and upper right back (arrows) and contusions. The contusions on the chest have a crisscross pattern. (c, d) Three-dimensional and coronal maximum intensity projection images show multiple anterior left rib fractures (arrows), dislocation of the left sternoclavicular joint, and a left scapular fracture.

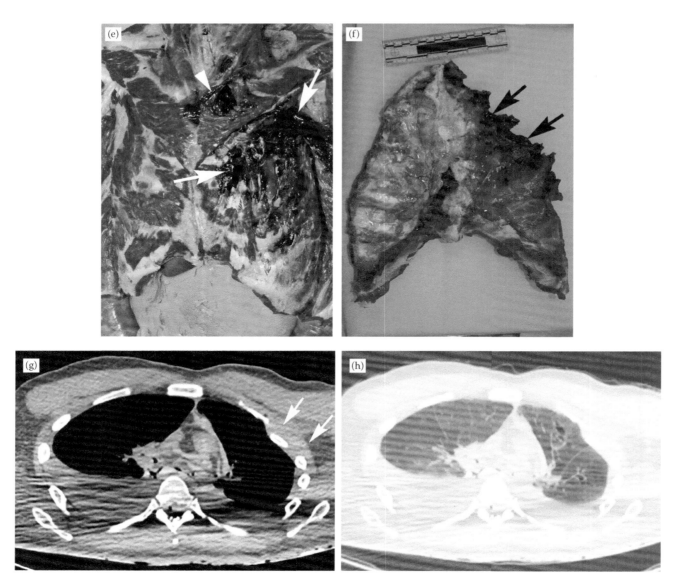

FIGURE 5.7 (*Continued*) A 39-year-old male pedestrian was struck by a motor vehicle and died of fatal head and chest injuries. (e, f) Autopsy photographs of the chest wall and removed anterior ribs show hemorrhage at the left sternoclavicular joint (arrowhead in e) and in the soft tissues and musculature of the left chest wall (arrows in e). Hemorrhage is present at the site of the rib fractures (arrows in f). (g) Axial MDCT shows hemorrhage in the soft tissue adjacent to the left rib fractures (arrows), mediastinal and pericardial hematomas, and bilateral hemothoraxes. (h) Axial MDCT in lung window settings shows large bilateral hemothoraxes and cystic change in the left lung from lung contusions and lacerations.

capsule. Renal contusions and lacerations occur when blunt force is applied to the flanks (Figure 5.9).

Spine, Pelvic, and Extremity Injuries

Blunt force injury to the spine, pelvis, and extremities consists of abrasions, lacerations, contusions in the skin and soft tissues, fractures, avulsions, and amputations. Accidents that have excessive impact and force result in the most severe of these injuries. Motor vehicle accidents

with pedestrians and aviation accidents are often characterized by the most severe injuries, such as soft tissue avulsion and amputation. Death may occur at the time of injury from fatal hemorrhage from vascular laceration, or cord compression or transection from vertebral fractures. Delayed deaths may be the result of complications such as pulmonary embolism, fat embolism, or infection.

Direct or indirect force may fracture bones. Fractures from direct forces are those that occur when traumatic

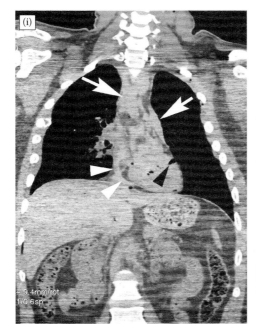 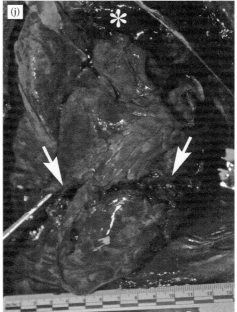

FIGURE 5.7 (*Continued*) A 39-year-old male pedestrian was struck by a motor vehicle and died of fatal head and chest injuries. (i) Coronal MDCT of the thorax and upper abdomen show mediastinal hemorrhage (arrows), hemorrhage in the right hilum, pericardial hemorrhage (white arrowheads), and gas in the left ventricular myocardium suggestive of a laceration (black arrowhead). (j) Autopsy photograph of the heart and mediastinum show mediastinal hemorrhage (asterisk) and transection of heart through both ventricles and intraventricular septum (arrows). Lacerations of the ascending aorta and pulmonary artery were also found.

forces are directly applied to a specific anatomic area. Indirect forces fracture bones that are located at an anatomic site remote from the site of applied force (Di Maio and Di Maio 2001). Indirect forces are usually the cause of vertebral fractures. At autopsy, the type of vertebral fracture can help establish the direction of force and mechanism of injury. The presence or absence of associated injury to the spinal cord is important in establishing the fracture as the lethal injury.

Pelvic fractures are an important component of blunt trauma because they indicate that a substantial amount of force has been applied to the body. Furthermore, pelvic fractures may cause death when there is arterial and venous injury. In such cases, massive pelvic and retroperitoneal hemorrhage may be the only finding at autopsy.

Nonfatal blunt force injuries to the extremities are often helpful in determining the mechanism of death or the circumstances of death. Defense wounds in a homicide victim are an example of the latter. Abrasions, contusions, and lacerations on the ulnar aspect of the forearms, backs of the hands and wrists, and over the knuckles are classic defensive wounds. In some cases, it may be possible to match the weapon with the wound pattern (Di Maio and Di Maio 2001, Murphy 1991).

RADIOLOGIC PRINCIPLES
Goals of Imaging

Postmortem multidetector computed tomography (MDCT) is useful to visualize and reconstruct blunt injury patterns prior to autopsy (Donchin et al. 1994, Jacobsen et al. 2008). In some cases, multiplanar and volumetric reformatted MDCT images may provide better visualization of blunt traumatic injuries than autopsy. Three-dimensional display of head, spine, and pelvic injuries may facilitate the understanding of the mechanism of injury.

Imaging Findings

Craniocerebral Injuries — Scalp lacerations and subgaleal hematomas cause focal soft tissue swelling on MDCT (Figure 5.1). The specific site of scalp laceration may be difficult to initially identify on MDCT. However, the asymmetry and increase in soft tissue attenuation of the scalp which result from the associated contusion and hemorrhage may allow the site of injury to be identified

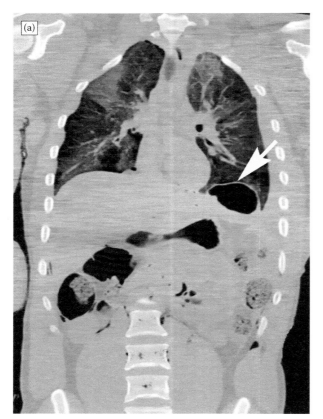
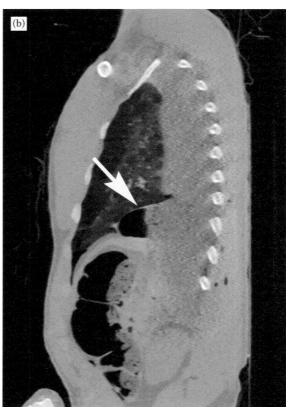

FIGURE 5.8   Motor vehicle accident victim. (a, b) Coronal and sagittal MDCT of the torso shown in lung window shows a traumatic diaphragmatic hernia on the left (arrow) and bilateral pulmonary opacities that were shown to be pulmonary contusions at autopsy. The sagittal image also shows a left hemothorax.

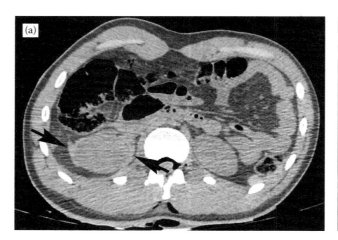
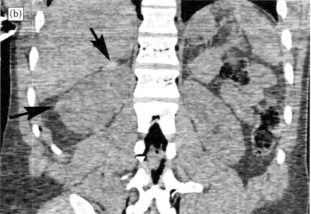

FIGURE 5.9   A 35-year-old man sustained blunt trauma to the flank in a motor vehicle accident. (a, b) Axial and coronal MDCT shows high-attenuation perinephric hemorrhage (arrows). At autopsy, the right renal vein was lacerated.

on MDCT if the edges of the laceration are open. Closed lacerations or those located on the dependent surfaces of the body are usually not visible on MDCT. Three-dimensional surface-rendering algorithms may be helpful in some cases.

Nondisplaced linear skull fractures appear as linear lucencies in the skull and usually involve both the inner and outer tables of the skull (Figure 5.2). The margins of linear skull fractures are well defined, lack sclerosis, and may cross sutures and vascular impressions. Depressed skull fractures have fragments that are displaced inward toward the brain (Figure 5.3). Three-dimensional volume-rendered MDCT images are very helpful to display the skull fracture pattern, which is often more difficult to appreciate at autopsy because the fracture fragments tend to fall apart when the scalp is peeled away from the calvarium. In catastrophic calvarial fractures in which the skull is crushed, a repeat MDCT after the medical examiner partially or completely reassembles the skull may be useful for analyzing the characteristics of the force that was applied to the skull.

Epidural hematomas are located between the skull and dura. A skull fracture with tear of the middle meningeal artery or dural venous sinus is present in 85% to 95% of epidural hematomas (Osborn 1994). Epidural hematomas are typically biconvex in shape and have mass effect on the adjacent brain. They may cause death by producing mass and cerebral herniation. Acute epidural hematomas are classically hyperattenuating on MDCT. However, they may be more heterogeneous when multiple bleeding episodes occurred prior to death or if there is decomposition. Putrefactive gas may be present within the hematoma.

Cerebral contusions occur with or without an associated skull fracture. In closed head trauma, contusions are more common on the gyral crests adjacent to bony protuberances in the skull. On MDCT, they appear as focal punctate or linear areas of hyperattenuating hemorrhage (Figure 5.2). The temporal tips, the cortex around the sylvian fissure, the inferior surface of the frontal lobes, and the parasagittal regions of the convexities are the most common locations. In our experience, small cerebral contusions are very subtle and difficult to identify on postmortem MDCT. Surrounding low-attenuation edema may be present if the decedent survived for a period of time after trauma.

Subdural hematomas are located between the dura and the arachnoid membrane. They are crescent shaped and do not cross dural attachments (Figure 5.5). Acutely, they are hyperattenuating on MDCT but may also be mixed attenuation. Chronic subdurals are typically fluid attenuation on MDCT, because they are composed of serosanguinous fluid. Decomposition may alter their appearance. We have found it difficult to detect small subdural hematomas that are thinly layered beneath the dura because on postmortem MDCT, the dura appears denser than the adjacent

brain, and the relative density of adjacent blood is similar to that of the dura.

Diffuse axonal injury classically occurs in the corticomedullary junction of the lobar white matter, corpus callosum, and dorsolateral aspect of the brain stem. MDCT may be normal or show petechial hemorrhages in the corpus callosum and at the gray–white junction. Postmortem magnetic resonance imaging (MRI) may be more effective at demonstrating diffuse axonal injury than MDCT; however, the findings of diffuse axonal injury on MRI have not been reported to date.

Subarachnoid hemorrhage is present in most cases of moderate to severe head trauma (Osborn 1994). It is a thin layer of high attenuation in the cerebrospinal fluid spaces, cisterns, and sulci on MDCT (Figure 5.6). In our experience, it is often difficult to correlate subtle areas of hemorrhage that are suspected on MDCT with autopsy because blood enters the subarachnoid space during removal of the calvarium. Decomposition makes the diagnosis of subarachnoid hemorrhage more challenging because the dura adjacent to the brain appears relatively dense as decomposition begins to occur. In addition, blood decreases in attenuation as it decomposes.

Vascular injuries to the carotid and vertebral arteries are difficult to diagnose on routine postmortem MDCT. Hemorrhage in the adjacent tissues is suggestive of an underlying laceration, but the location and extent of laceration are not detectable on cross-sectional imaging unless intravascular contrast is administered. Similarly, intravascular contrast is necessary to diagnose dissections on MDCT. There is great potential for MDCT angiography to augment autopsy in these cases, because dissection and evaluation of the carotid and vertebral basilar systems can be time consuming and difficult.

Thoracoabdominal Injuries

Preautopsy imaging in blunt chest trauma is useful to show pneumothorax, tension pneumothorax, and the placement of tubes and lines if resuscitation was attempted. Clinically significant pneumothorax can usually be distinguished from early decompositional gas in the pleural space. Early decompositional gas in the pleural space is usually small in volume and accompanied by decompositional gas in other locations, such as in the visceral vessels of the abdomen, great vessels, and heart. Furthermore, decompositional gas usually has a symmetric distribution in the body. The presence of an associated rib fracture, hemothorax, and pulmonary contusion supports the diagnosis of traumatic pneumothorax. A small occult traumatic pneumothorax without an associated injury may be difficult to distinguish from early decompositional gas.

Pulmonary contusions most often occur at the site of impact. Airspace consolidation and opacification in a nonsegmental distribution is a characteristic finding (Figure 5.8) (Aghayev et al. 2008). Consolidation in the contralateral portion of the chest is indicative of a contracoup contusion. Lucency between the consolidation and pleural surface of the lung may be observed (Kaewlai et al. 2008). Pulmonary lacerations may appear as focal consolidations or cavities on MDCT (Figure 5.7). They may have surrounding opacity from contusion. Linear tracks of gas through the lung may also indicate communication with a bronchus and an associated tracheal or bronchial laceration. Tracheal or bronchial lacerations may also produce pneumomediastinum.

Hemorrhage in the mediastinum is indicative of a major vascular injury. Aortic lacerations are one of the most common major vascular injuries in blunt trauma. Radiography may show widening of the mediastinum from a periaortic hematoma, blurring of the aortic contour, or thickening of the paratracheal stripe. In some cases, radiography is normal. On MDCT, mediastinal hematoma is the most indicative finding of aortic laceration (Figure 5.7). The position and contour of the aorta may be altered, but this is difficult to appreciate on postmortem imaging because the aorta is frequently collapsed because of intravascular volume loss and loss of systemic blood pressure. MDCT angiography is potentially useful to identify the site of rupture. Injuries to the aortic arch branches, pulmonary artery, and vena cava may also produce mediastinal hematomas. Pericardial and cardiac lacerations usually result in pericardial hematomas, which may cause cardiac tamponade. Cardiac contusions and lacerations are usually not evident on postmortem MDCT (Aghayev et al. 2008). In the absence of decomposition, gas collections in the myocardium should raise suspicion for laceration (Figure 5.7i).

Diaphragm elevation should raise concern for diaphragm laceration or rupture. Intra-abdominal organs may protrude into the thorax when there is laceration or rupture of the hemidiaphragm (Figure 5.8). Laceration or rupture of the liver, spleen, and other visceral organs may be difficult to identify on routine postmortem MDCT because of the noncontrast technique. Furthermore, when a trauma victim dies of multisystem injury, the amount of hemorrhage in the abdomen may be less than expected if the victim has had massive hemorrhage from great vessel or head injury. Hemoperitoneum usually has a higher attenuation than simple ascites, which is water attenuation. However, if there is decomposition, the attenuation of intraperitoneal or retroperitoneal blood may be lower than expected. Focal collections of blood adjacent to an organ or major vascular structure are indicative of injury (Figure 5.9). The site of injury may not specifically be identifiable on MDCT. Extraluminal gas within the abdomen is very commonly observed on postmortem MDCT because the earliest signs of decomposition are observed in the abdomen. Therefore, the presence of pneumatosis, intravascular gas, pneumoperitoneum, and pneumoretroperitoneum should be interpreted with caution. Asymmetric or focal collections of gas should be viewed with suspicion, and the surrounding organs should be carefully evaluated to try to determine if the etiology of the gas can be explained by organ injury rather than decomposition.

Spine, Pelvic, and Extremity Injuries

Direct trauma to the spine, pelvis, and extremities is usually accompanied by soft tissue trauma, which is better identified and evaluated with physical examination than imaging. Abrasions, contusions, and minor hemorrhages into the soft tissues may not be evident on postmortem MDCT. Significant hemorrhage into the soft tissues increases the attenuation and thickness of the involved soft tissue. With increasing hemorrhage into the soft tissues, the fat planes become distorted, and focal asymmetry develops.

Diagnosis and interpretation of fractures are generally straightforward on radiography. Fractures are linear, angulated, or displaced lucencies within bone. Vertebral body compression fractures can be identified when there is loss of vertebral body height or increased density within the bone from the compressive forces. Vertebral body compression fractures and abnormalities in alignment are best viewed on sagittal MDCT images (Figure 5.10). Axial images are useful to view the pedicles and posterior elements of the vertebral bodies. Intervertebral disc injuries and spinal cord contusion or hematoma cannot be reliably assessed on MDCT. We have found MDCT to be very useful to screen the bony structures of the spine, particularly the cervical spine, because this region is difficult to dissect at autopsy. Three-dimensional images provide an excellent depiction of the anatomic distribution of spine fractures, which can be difficult to appreciate at autopsy (Figures 5.10 and 5.11).

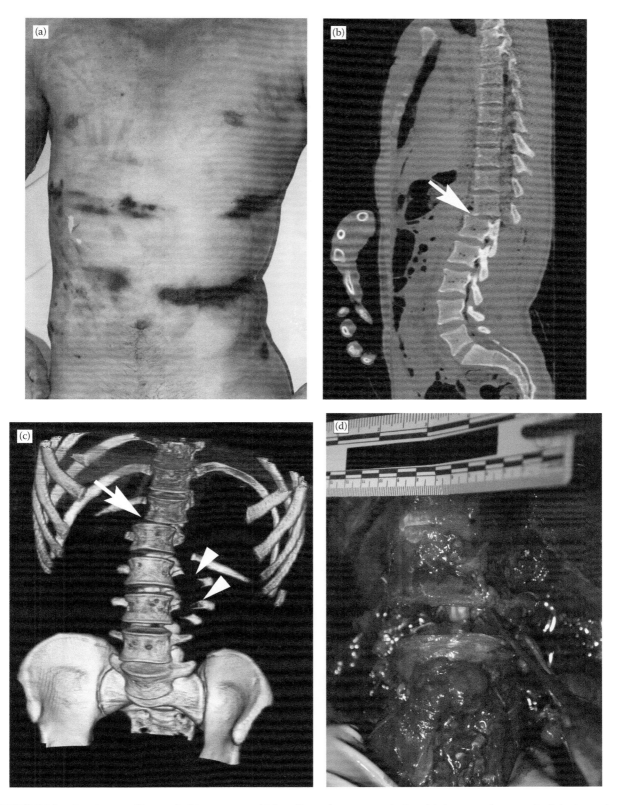

FIGURE 5.10   An 18-year-old man died in a motor vehicle rollover due to an explosion. (a) Autopsy photograph of the torso shows contusions and abrasions on the lower chest and upper abdomen. (b, c) Sagittal and three-dimensional MDCT images of the torso show fracture dislocation of the spine at the T12-L1 level (arrows) and multiple transverse process fractures (arrowheads in c). (d) Autopsy photograph of the dissected spine shows transection of the spinal cord.

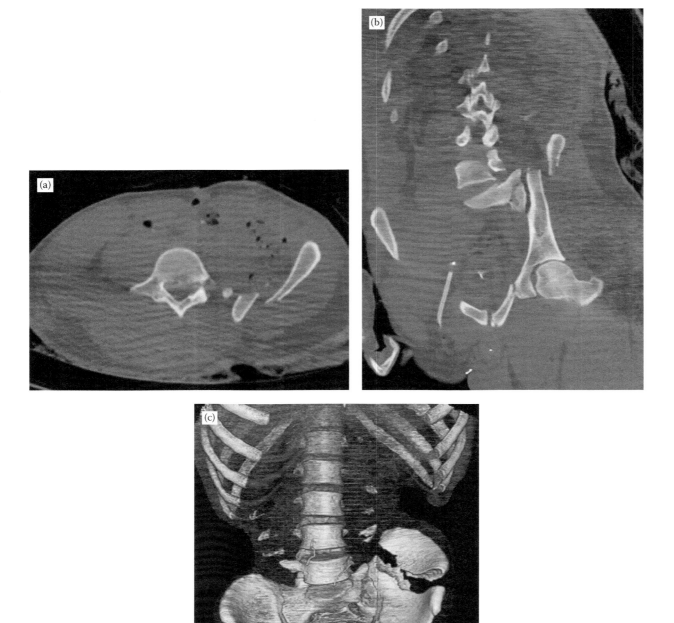

FIGURE 5.11   A 25-year-old man died from multisystem trauma in a motor vehicle rollover. (a, b) Axial and oblique coronal MDCT show a displaced fracture of the left iliac crest. (c) Three-dimensional MDCT shows the displaced left iliac fracture, multiple bilateral transverse process fractures, and a left acetabular fracture. A Foley catheter and a right femoral catheter are in place from the resuscitation attempt.

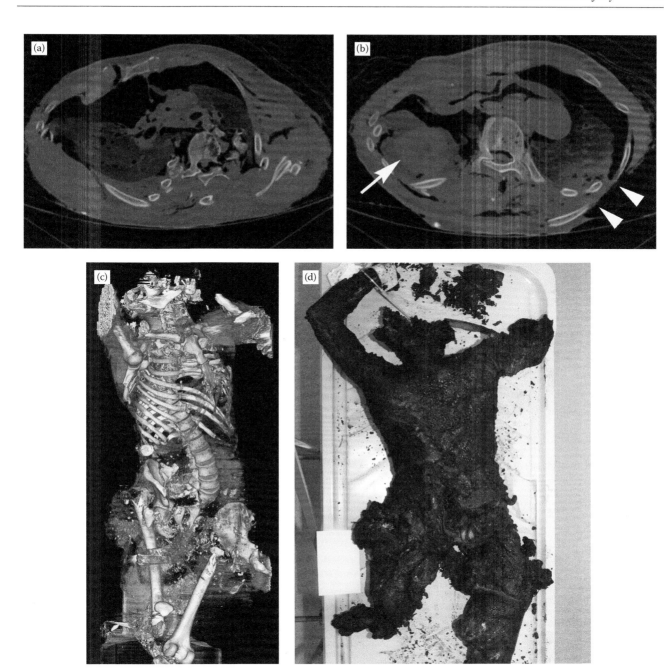

FIGURE 5.12 Extensive blunt traumatic and thermal injury to the entire body in an adult male victim of a helicopter crash. (a, b) Axial MDCT of the chest shows crush injury to the chest that deforms the thoracic cavity from fractures of the ribs and spine. There is disruption of the mediastinal and cardiac structures and herniation of the right kidney (arrow) into the chest through a diaphragmatic laceration. Thermal tissue loss (arrowheads) is also present. (c) Three-dimensional MDCT of the torso shows innumerable skeletal fractures, spinal fracture dislocation, crush injury of the pelvis, and amputation. (d) Autopsy photograph shows external thermal injury and amputations of the cranial vault and extremities.

Complex, comminuted, and open fractures of the extremities are generally simple to diagnose. Fractures of the extremities, hands, and feet are best evaluated with conventional radiography. However, MDCT has the added benefit of providing soft tissue information that may be valuable in the assessment of the extent and volume of an associated hematoma, particularly when there is trauma in the proximal extremities such as the thigh and injury to the femoral vessels is suspected. As previously discussed, the precise location and extent of vascular laceration are not detectable on MDCT unless intravascular contrast is administered. Three-dimensional MDCT is very helpful to visualize the entire injury pattern if analysis of the injury mechanism is necessary.

Blunt force injuries may also coexist with other injury mechanisms. For example, explosions may occur when combustible fuel is ignited in aviation and motor vehicle accidents. The victims of these accidents may have a combination of blunt force, blast, and thermal injury (Figure 5.12). The features of blast and thermal injuries are discussed in later chapters.

CONCLUSIONS

Postmortem MDCT in blunt trauma provides the medical examiner with information that saves time and increases the accuracy of the autopsy. In some cases, it may enable the medical examiner to focus the autopsy dissection on a particular anatomic region that is likely the site of major injury that caused death. Multiplanar two- and three-dimensional MDCT images are very useful to depict complex skull and skeletal trauma that may be difficult to appreciate at autopsy. The fracture patterns may be indicative of the position of the victim at the time of death and may aid in interpretation of the direction of the blunt force. MDCT provides a very limited assessment of vascular and solid organ injury; however, MDCT angiography may be

helpful to elucidate vascular injury if a limited autopsy is desired.

REFERENCES

Adams, J. H., Doyle, D., Ford, I. et al. 1989. Diffuse axonal injury in head injury: definition, diagnosis and grading. *Histopathology* 15: 49–59.

Aghayev, E., Christe, A., Sonnenschein, M. et al. 2008. Postmortem imaging of blunt chest trauma using CT and MRI: comparison with autopsy. *J Thorac Imaging* 23: 20–27.

Davis, G. J. 1998. Patterns of injury. Blunt and sharp. *Clin Lab Med* 18: 339–350.

Di Maio, V. J. M., and Di Maio, D. J. 2001. *Forensic pathology*, Boca Raton, FL: CRC Press.

Donchin, Y., Rivkind, A. I., Bar-Ziv, J. et al. 1994. Utility of postmortem computed tomography in trauma victims. *J Trauma* 37: 552–555; discussion 555–556.

Graham, D. I., Adams, J. H., Nicoll, J. A., Maxwell, W. L., and Gennarelli, T. A. 1995. The nature, distribution and causes of traumatic brain injury. *Brain Pathol* 5: 397–406.

Jacobsen, C., Schon, C. A., Kneubuehl, B., Thali, M. J., and Aghayev, E. 2008. Unusually extensive head trauma in a hydraulic elevator accident: post-mortem MSCT findings, autopsy results and scene reconstruction. *J Forensic Leg Med* 15: 462–466.

Kaewlai, R., Avery, L. L., Asrani, A. V., and Novelline, R. A. 2008. Multidetector CT of blunt thoracic trauma. *Radiographics* 28: 1555–1570.

Lederer, W., Mair, D., Rabl, W., and Baubin, M. 2004. Frequency of rib and sternum fractures associated with out-of-hospital cardiopulmonary resuscitation is underestimated by conventional chest X-ray. *Resuscitation* 60: 157–162.

Morrison, A. L., King, T. M., Korell, M. A., Smialek, J. E., and Troncoso, J. C. 1998. Acceleration-deceleration injuries to the brain in blunt force trauma. *Am J Forensic Med Pathol* 19: 109–112.

Murphy, G. K. 1991. "Beaten to death." An autopsy series of homicidal blunt force injuries. *Am J Forensic Med Pathol* 12: 98–101.

Osborn, A. G. 1994. *Diagnostic neuroradiology*, St. Louis: Mosby.

Swierzewski, M. J., Feliciano, D. V., Lillis, R. P., Illig, K. A., and States, J. D. 1994. Deaths from motor vehicle crashes: patterns of injury in restrained and unrestrained victims. *J Trauma* 37: 404–407.

Wiegmann, D. A., and Taneja, N. 2003. Analysis of injuries among pilots involved in fatal general aviation airplane accidents. *Accid Anal Prev* 35: 571–577.

Chapter 6

Death from Fire and Burns

FORENSIC PRINCIPLES

In the United States, more than 3400 civilians die as a result of fire each year (Administration 2007). More people die in fires than all natural disasters. Fire death investigation is challenging because antemortem and postmortem burns are indistinguishable at autopsy and not all victims recovered from a fire have died from the effects of fire. Death may have occurred from natural disease or injury before the fire or as a result of the fire. Most deaths associated with fires are accidental; however, suicide or homicide before the fire, and homicide by the fire should always be considered when investigating fire deaths.

Classification and Prognosis of Burns

Burns are tissue injuries caused by extreme heat, flame, contact with heated objects, or chemicals. The extent and severity of a burn are dependent on the temperature of the heat, time of exposure, and presence of clothing or covering on the skin. The most commonly used classification for burns reflects the depth of skin affected: partial-thickness or full-thickness burns. Partial-thickness burns, formerly called superficial, first-degree, and second-degree burns, are limited to the epidermis and do not extend into dermal tissues. In contrast, full-thickness burns, formerly called deep, third-degree, and fourth-degree burns, destroy both the epidermis and dermis (Kane and Kumar 2005). Charring of fat, muscle, and bone may occur in full-thickness burns.

The prognosis of burn victims is related to the age, amount of body surface area burned, and inhalation injury (Ryan et al. 1998). Victims who are over the age of 60, have more than 40% of the body-surface area burned, or have smoke inhalation injury are more likely to die from their burns. Immediate death from a fire is due to direct thermal injury or smoke inhalation. In contrast, delayed deaths occur within the first three days after injury and are due to shock, fluid loss, or respiratory failure from inhalation injury (Di Maio and Di Maio 2001).

Autopsy Findings

Determining the cause and manner of death for a victim found dead at the scene of a fire may be difficult and require careful consideration of data obtained from the scene of the fire, historical information, and circumstantial evidence with the autopsy findings. A body found at the scene of a fire may appear normal without any evidence of injury; show evidence of prolonged exposure to heat with darkening and stiffening of the skin; have blisters and hyperemia from partial-thickness burns (Figures 6.1 and 6.2); or have full-thickness burns with a variable degree of tissue loss, destruction, and charring (Figure 6.3).

Partial-thickness burns produce moist blisters with surrounding hyperemia. The finding of hyperemia surrounding a burn is a useful finding that suggests the injury occurred during life, unless death occurred very shortly after the injury and vital reaction did not have time to develop. Skin slippage may also be apparent in partial-thickness burns (Figure 6.1). Full-thickness burns show tissue coagulation and may be brown or black from charring (Figures 6.2 and 6.3).

Fire victims who have no evidence of burns or have minimal thermal injury may have died before the fire or during the fire. In the latter circumstance, death may occur from heat, inhalation of toxic gases, or physical trauma sustained trying to escape the fire or from structural collapse. Death from the effects of heat (circulatory collapse) can occur without thermal damage to the skin or significant elevation of serum carboxyhemoglobin (COHb) or cyanide (Hill 1989, Lawler 1993).

The composition and volume of toxic gases in a fire depend on the nature of material burning and temperature of the fire. Although there are numerous highly toxic gases generated by combustion, carbon monoxide and hydrogen cyanide are the most commonly analyzed in death investigation. The presence of soot in the pharynx and lower airways along with cherry red coloration of livor mortis, muscles, blood, and organs is suggestive of the decedent being alive and breathing during the fire. Many will have elevated carbon monoxide saturation (Figure 6.4). Elevation of the percentage of serum COHb on toxicological examination supports these findings. Serum COHb provides an assessment of whether or not there was active

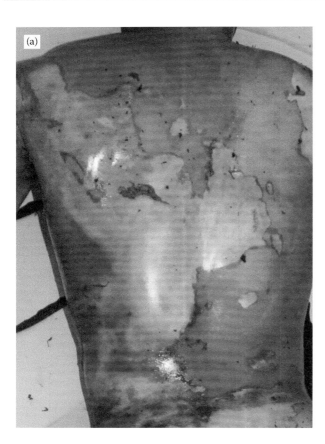

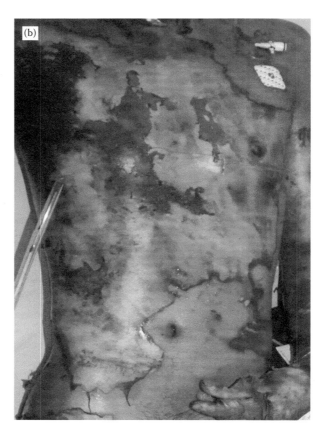

FIGURE 6.1   Partial- and full-thickness burns with some charring on a man who died in an explosion with fire. Autopsy photographs show multiple partial-thickness burns on the back (a) and chest (b). A right chest tube and left subclavian central line are in place from an attempt at resuscitation.

respiration during the fire and, thereby, aids in the determination of whether death occurred before or during the fire. Active respiration does not always correlate with the level of consciousness during the fire. COHb exceeding 3% in nonsmokers and 10% in cigarette smokers is considered abnormal (Ernst and Zibrak 1998). Over half of all deaths that occur during fire have a COHb exceeding 50% (Lawler 1993). However, lesser levels are often encountered (20% to 40%) because there may be contributory factors such as underlying cardiac or pulmonary disease, drug or alcohol intoxication, insufficient oxygen, and other toxic gases that may cause a person to die from lower COHb levels. Deaths associated with exhaust of an automobile may have little or no elevation of carbon monoxide saturation due to the conversion to carbon dioxide by modern catalytic converters.

In severely burned bodies with extensive charring, the physical autopsy may be limited. An understanding of the

changes that occur in a body as a result of fire is useful because postmortem thermal changes may mimic antemortem injury. When the body is exposed to high temperatures, skin and soft tissues rapidly contract and split (Figure 6.3b). This phenomenon produces *heat ruptures*, which appear as linear lacerations of the skin and underlying soft tissue (Lawler 1993). Characteristically, heat ruptures can be distinguished from antemortem lacerations because they do not show surrounding edema, erythema, or bruising. Continued exposure to fire results in thermal destruction of skin and subcutaneous tissue which exposes skeletal muscle, bone, and internal organs. The craniofacial region and the anatomic region most directly exposed to the fire often have the most pronounced soft tissue changes.

Skeletal muscles contract and shrink from the heat, resulting in flexion deformities in the extremities (Figure 6.5). When the upper extremities are flexed

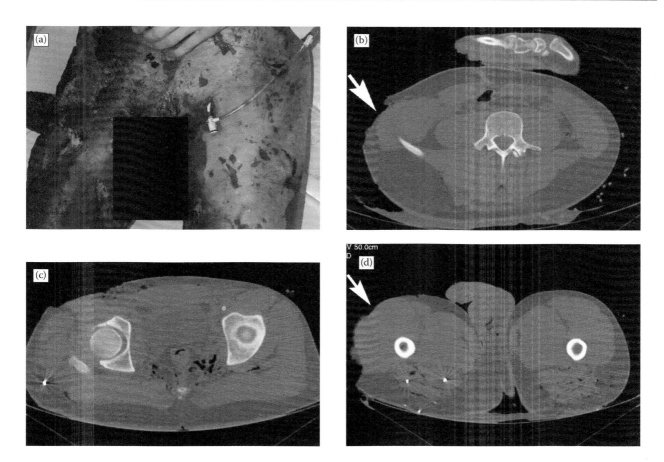

FIGURE 6.2 Partial- and full-thickness burns on a man who died in an explosion with fire. (a) Autopsy photograph shows multiple partial-thickness burns on the left upper thigh and full-thickness burns with charring and muscle exposure on the right lateral thigh. A left femoral catheter is in place from an attempt at resuscitation. (b, c, d) Axial MDCT images show absent dermis and subcutaneous fat on the right pelvis and anterior right thigh (arrows) corresponding to the full-thickness burns. The exposed muscle has an irregular surface contour. There are multiple metallic fragments within the posterior thigh and lateral to the right hip from projectiles related to the explosion and the left femoral catheter is seen in (c). Scattered collections of decompositional gas are present.

from thermal muscle contraction, the term *pugilistic attitude* is applied because the upper body position is that of a boxer holding his hands in front of him (Figure 6.6) (Di Maio and Di Maio 2001). Fine, linear, superficial fractures of the cortical surface of bone are characteristic of heat-induced fractures (Bohnert et al. 1998) (Figure 6.7). Thermal amputations occur as the fire consumes the distal extremities. Thermal amputations may be difficult to distinguish from traumatic amputations (Figure 6.8). The findings of rounded, burnished edges of the bone generally indicate a thermal amputation, whereas sharp, well-demarcated edges usually indicate a traumatic amputation.

In the skull, thermal changes advance from fine, cortical fractures in the outer table, to loss of the outer table and development of a thermal epidural hematoma, to eventual progressive destruction of the cranium from the vertex to the skull base (Figure 6.9 through Figure 6.12) (Bohnert et al. 1997, Bohnert et al. 1998, Di Maio and Di Maio 2001). Thermal epidural hematomas are common findings and represent the accumulation of blood in the epidural space as the brain dehydrates and shrinks from heat. The blood in the hematoma is typically soft and friable, light brown or reddish, and has a honeycomb appearance (Figure 6.13). Although the blood is thought to originate from the dural venous sinuses and emissary veins, the exact mechanism of formation is unknown (Lawler 1993). Thermal epidural hematomas may look similar to traumatic epidural hematomas but are not associated with traumatic skull fractures or other cranial injuries (Figure 6.14).

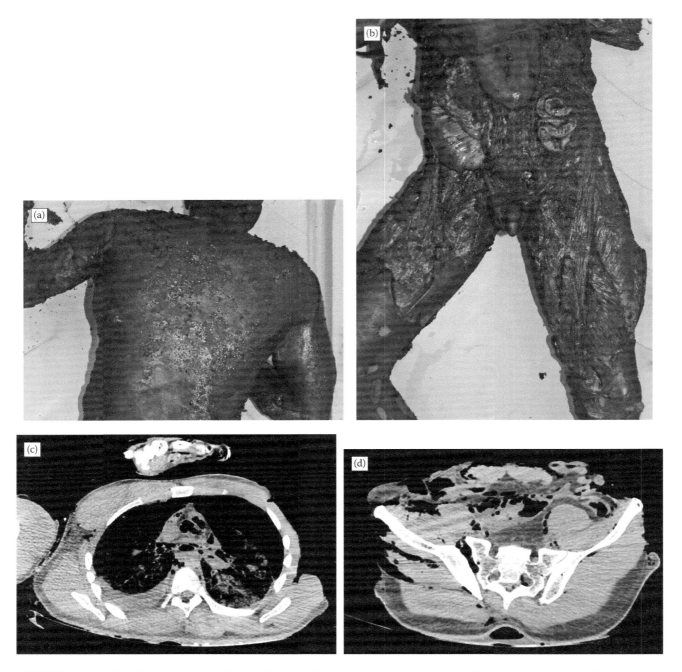

FIGURE 6.3   Full-thickness burns and charring from a vehicle crash with subsequent fire. (a) Autopsy photograph of a fire victim's back shows brown and black charring with tissue coagulation. (b) Autopsy photograph of the anterior abdomen, pelvis, and thighs shows heat rupture of the anterior abdominal wall with evisceration. Full-thickness burns on the thighs expose skeletal muscles. (c) Axial MDCT image of the chest shows thermal tissue loss with full-thickness burns of the chest wall and increased attenuation within the lungs. A small right hemothorax is present and was secondary to antemortem traumatic rib fractures and blunt chest trauma. (d) Axial MDCT of the upper pelvis shows near complete thermal tissue loss of the anterior abdominal wall and evisceration. There is diastasis of the right sacroiliac joint from blunt trauma sustained prior to the fire.

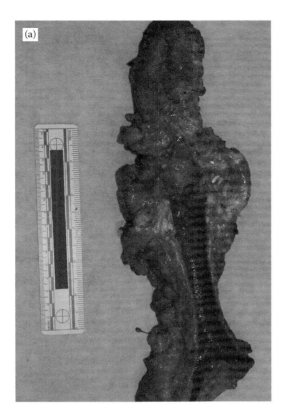

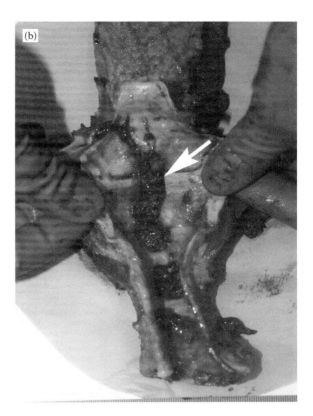

FIGURE 6.4 Autopsy findings of smoke inhalation in fire victims. (a) Autopsy photograph of the tongue and airways shows cherry red coloration of the soft tissues from carbon monoxide intoxication, searing of the tongue, and soot in the trachea. (b) Autopsy photograph from a different person than in (a) shows brown soot (arrow) in the subglottic trachea.

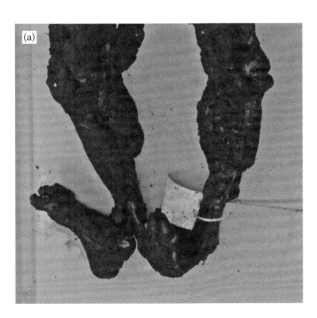

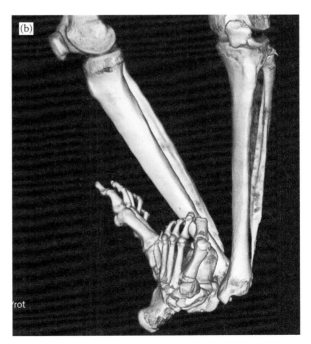

FIGURE 6.5 Flexion deformities and amputations in severe burns. (a) Autopsy photograph shows charred lower extremities and dorsiflexion of the ankles. (b) Three-dimensional volume rendering of the computed tomography data shows dorsiflexion of the ankles and plantarflexion deformities of the left toes. There is bilateral ankle disarticulation that is likely secondary to muscle retraction and flexion.

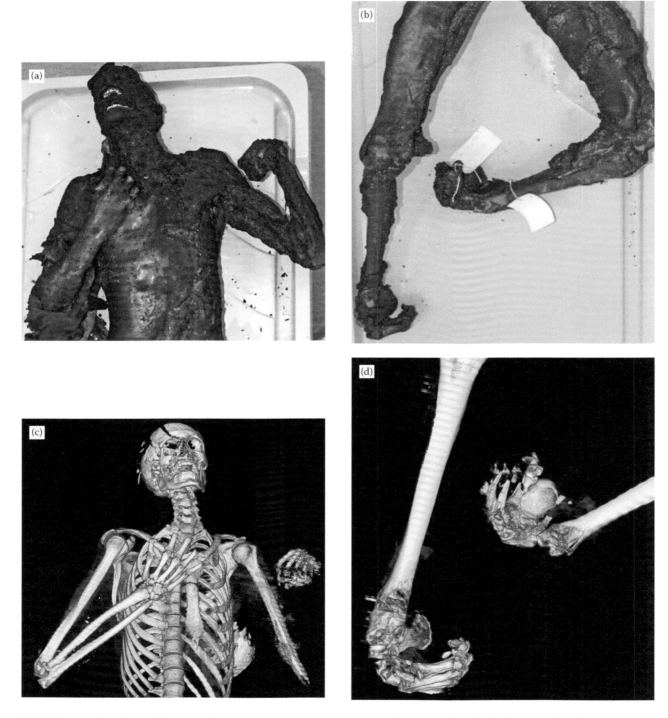

FIGURE 6.6 Pugilistic attitude. (a, b) Autopsy photograph shows severe charring and pugilistic flexion deformities of the upper extremities and lower legs. (c, d) Three-dimensional MDCT shows the flexion deformities of the upper extremities and lower extremities with associated fracture and amputation of the digits. The scan also reveals skull fractures that were sustained prior to the fire. The cause of death was blunt force injury to the head.

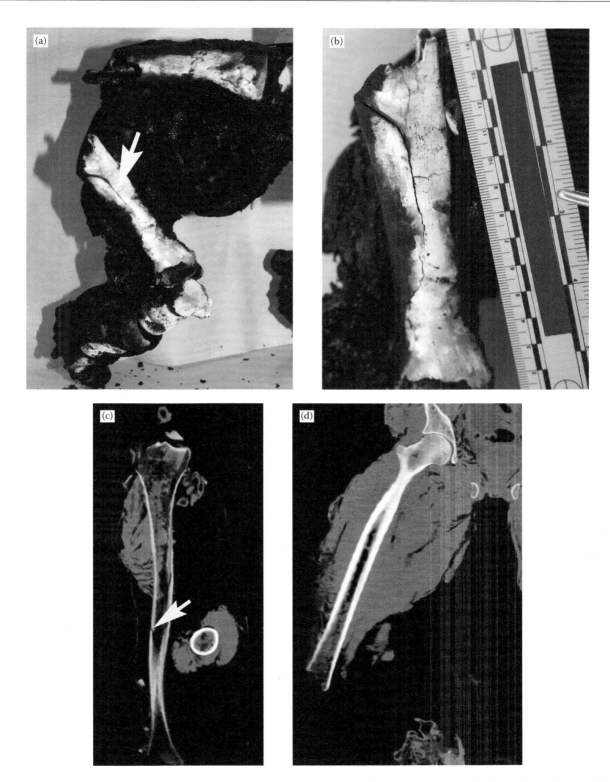

FIGURE 6.7   Spectrum of findings in thermal fracture and amputation. (a, b): Autopsy photographs show calcined bone and linear thermal fractures (arrow). The knee is absent in (a) from thermal destruction, and the distal femur and proximal tibia have the characteristic burnished margins of thermal amputations. (c, d) Coronal MDCT of the tibia and femur in a different person than in (a) and (b) show fine, linear thermal fractures (arrow in c), heat-related changes in the marrow spaces, and skeletal muscle contracture with thermal amputation of the distal femur in (d).

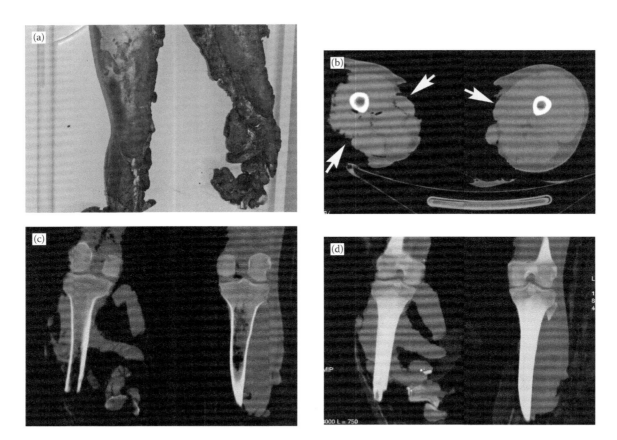

FIGURE 6.8   Thermal amputation and fractures in a charred body. (a) Autopsy photograph shows charring of both lower extremities with distal amputations. (b) Axial MDCT shows full-thickness burns of both lower extremities (arrows). (c, d): Bilateral thermal amputations of the tibia in a charred fire victim show mottled lucency and uncovered amputation margins on the coronal MDCT shown in (c). The maximum intensity projection image shown in (d) obscures the finding of mottled lucency but shows the smooth margin of the thermal amputations.

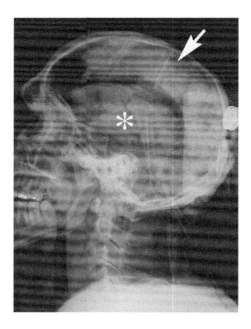

FIGURE 6.9   Thermal epidural hematoma on a lateral radiograph of the skull. The thermal epidural hematoma (arrow) is separate from the retracted brain parenchyma (asterisk).

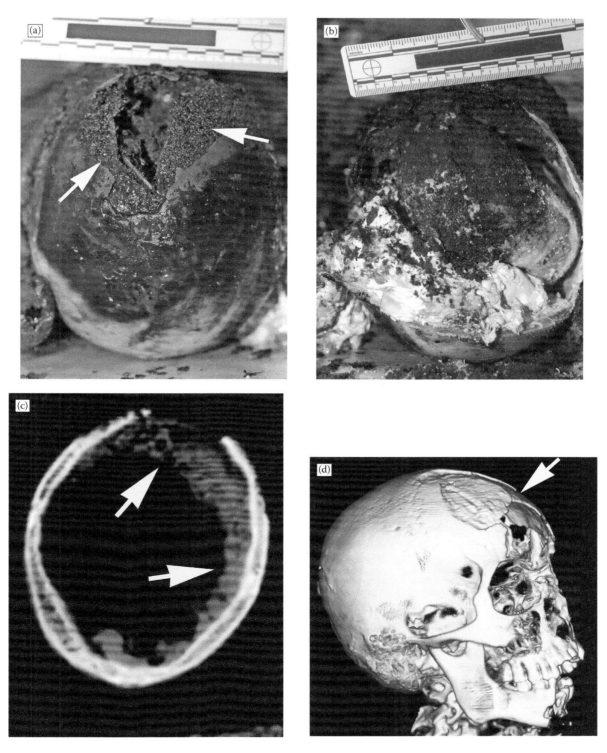

FIGURE 6.10 Thermal cranial findings in a fire victim who sustained antemortem traumatic facial fractures. (a, b) Autopsy photograph of the skull shows loss of the outer table (arrows) adjacent to the thermal cranial defect. The opened skull in (b) shows retracted brain and reddish brown thermal epidural hematoma. Note the autopsy photographs are reversed to correspond with the anatomic orientation of the MDCT images. (c) Axial MDCT image of the skull shows the classic honeycomb pattern of a thermal epidural hematoma (arrows), thermal loss of the skull and outer table, and mottled lucency in the marrow spaces. (d) Three-dimensional volume-rendered MDCT shows the traumatic mandibular fractures and thermal loss of the outer table of the skull (arrow) adjacent to a thermal cranial defect.

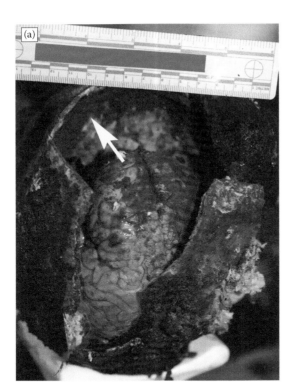

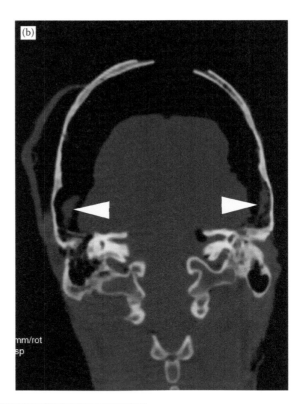

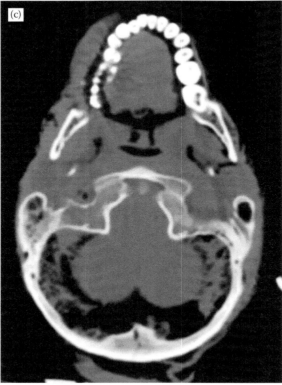

FIGURE 6.11   Thermal epidural hematoma. (a) Autopsy photograph shows a thermal defect in the calvarium, retracted brain, and thermal epidural hematoma (arrow). (b) Coronal MDCT shows retraction and loss of the scalp, a thermal defect in the calvarium, and retraction of the brain parenchyma. Small thermal epidural hematomas are present (arrowheads). (c) Axial MDCT through the posterior fossa shows a retracted cerebellum with adjacent thermal epidural hematoma. Note thermal tissue loss of the left face and scalp.

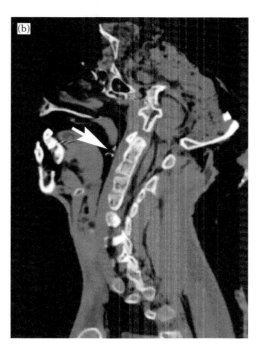

FIGURE 6.12   Severe craniofacial thermal destruction in a fire victim who had an elevated carboxyhemoglobin and autopsy evidence of inhalation injury. (a) Autopsy photograph of the posterior aspect of the head and neck shows near complete thermal destruction of the skull. There is a small amount of residual shrunken brain (arrows). (b) Sagittal MDCT image shows thermal calvarial and facial bone and soft tissue destruction with residual brain tissue in the posterior fossa as shown in the autopsy photograph. There is a focus of high attenuation in the oropharnyx that may represent debris (arrow).

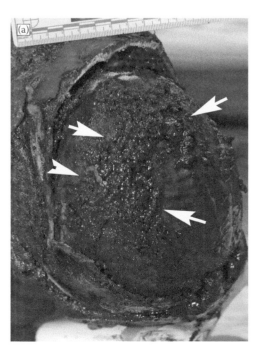
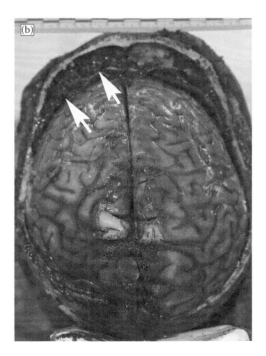

FIGURE 6.13   Thermal epidural hematoma without desiccation. (a, b) Autopsy photographs show a right-sided thermal epidural hematoma (arrows). Note that the autopsy photographs have been reversed to correspond with the anatomic orientation of the MDCT images.

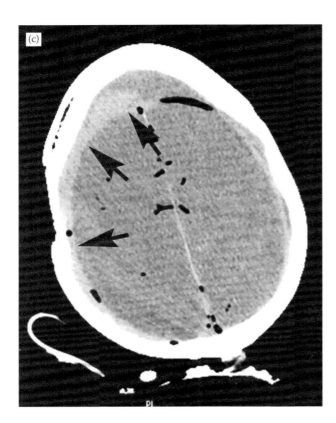
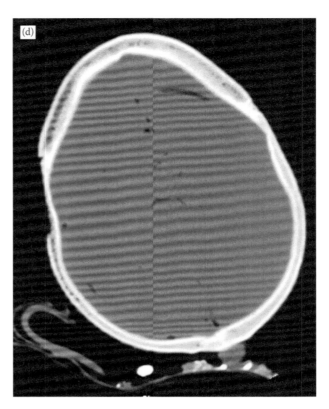

FIGURE 6.13 (*Continued*) Thermal epidural hematoma without desiccation. (c, d) Axial MDCT images show mixed high- and low-attenuation blood in the epidural space (arrows in c) corresponding to the autopsy findings. The same image viewed with a bone window setting (d) shows no evidence of a skull fracture. There is thermal loss of the outer table of the skull on the right.

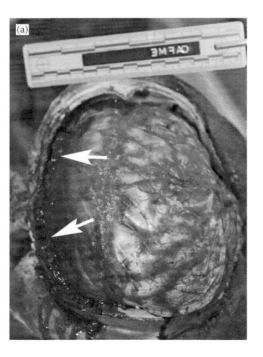
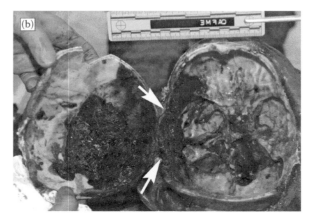

FIGURE 6.14 Antemortem traumatic epidural hematoma in a fire victim who died prior to the fire from severe head trauma. (a, b) Autopsy photographs of the opened skull show a right-sided epidural hematoma overlying the intact dura mater (arrows). Note that the autopsy photographs have been reversed to correspond with the anatomic orientation of the MDCT images.

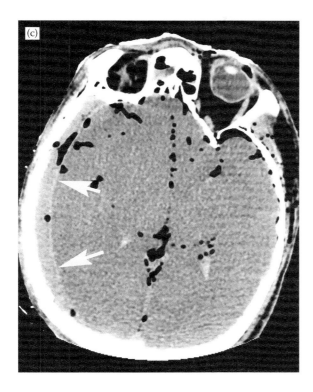 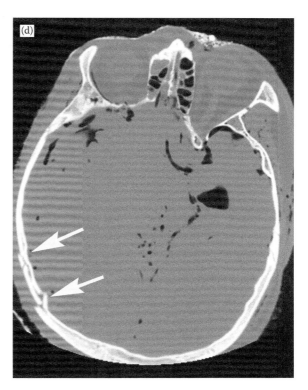

FIGURE 6.14 (*Continued*) Antemortem traumatic epidural hematoma in a fire victim who died prior to the fire from severe head trauma. (c, d) Axial MDCT shows a high-attenuation epidural hematoma (arrows in c) and associated traumatic skull fractures (arrows in d). A small amount of intraventricular blood is in the occipital horns of the lateral ventricles (seen in c) from the associated parenchymal brain injury.

RADIOLOGIC PRINCIPLES

Goals of Imaging

The primary goal of imaging a burned body prior to autopsy is to assist the medical examiner in excluding other causes of death. Multidetector computed tomography (MDCT) is particularly useful in severely burned and charred bodies that are difficult to examine. The more severely burned an individual becomes, the more problematic it is to establish the exact cause of death, because thermal changes may obscure the detection of antemortem injury or disease. MDCT accurately localizes and documents traumatic fractures and injuries that may have occurred before or during the fire (Levy et al. 2009). When the possibility of death before the fire is being considered, MDCT is very useful to localize bullet fragments or bone injury that may have been inflicted before the fire. MDCT may also aid in the identification of charred and fragmented remains through finding distinctive anatomic features and surgical hardware that can be compared to antemortem radiographs and computed tomography (CT) scans. Furthermore, MDCT may identify tissue suitable for DNA analysis (Harcke et al. 2009).

Imaging Findings

The physical findings of smoke inhalation are easily identified at autopsy and confirmed with serum COHb analysis. MDCT does not contribute to the diagnosis of smoke inhalation and COHb intoxication. Soot is not visualized on MDCT. However, particulate material and debris may occasionally be identified within the air passages as foci of high-attenuation particulate material in the airways (Figure 6.12b).

Partial-thickness burns may produce no significant changes in the dermis or mild irregularity of the dermis on MDCT. Full-thickness burns show loss of the dermal layer with exposure of the underlying fat or skeletal muscle. The outer margins of the exposed fat and muscle are typically irregular and jagged (Figures 6.2 and 6.3). The degree and extent of fat and muscle exposure varies with the severity of the burns. In severely charred victims, skeletal muscle is exposed and retracted. Retraction of skeletal muscle is due to shortening (or "cooking") of the muscle fibers from the heat of the fire. As the muscles shorten, adjacent joints flex, creating characteristic flexion deformities in the upper and lower extremities (Figures 6.5 and 6.6). Shortening and thermal destruction of muscle also causes the muscle to pull away from the distal ends of the bone such that soft tissue does not cover the distal ends of bone. This may be observed in the extremities as well as the torso (Figure 6.8 and Figure 6.15).

On MDCT, flexion deformities are most evident on three-dimensional volume-rendered images (Figures 6.5 and 6.6). Thermal flexion deformities are associated with fractures and dislocations from the mechanical forces associated with muscular contraction and shrinkage which create unbalanced forces on the joints. The sequence of flexion contracture and fracture/dislocation likely precedes the thermal destruction of bone that results in thermal amputations. The combination of retraction, flexion, dislocation, and fracture should facilitate the recognition that the findings are due to thermal injury rather than injury that occurred prior to death or before the fire (Levy et al. 2009).

Traumatic fractures and amputations in severely burned remains may cause interpretive difficulties, because heat-related cortical fractures and amputations are common in severely charred remains. Thermal fractures are linear cortical fractures in bone uncovered by soft tissue or bone (Figure 6.7). They are typically found in areas of severe charring and exposure to heat. In contrast, traumatic fractures are found in unexposed bone and are typical of mechanical injury such as a spinal compression fracture, complex pelvic fracture with involvement of the sacroiliac joints and symphysis pubis, and oblique or comminuted fractures of the distal extremities and ribs (Figure 6.16). Thermal amputations have smooth transverse or angulated margins that are not covered by skeletal muscle because of thermal-related skeletal muscle shrinkage and retraction. In contrast, traumatic amputations have sharp, angulated margins or evidence of comminution. Maximum intensity projection images and three-dimensional images are useful in the evaluation of amputations because they replicate the appearance of the bones at autopsy by showing the margins and contours of the ends of the bones (Figure 6.8). Also indicative of heat-related bone injury is the finding of mottled lucency in the marrow space on multiplanar two-dimensional images (Figures 6.7 and 6.8) (Levy et al.

2009). The lucencies represent thermal loss of the normal marrow elements and thin trabecula with preservation of the thicker bony trabecula.

The craniofacial region often shows the most dramatic findings in severely burned remains. Facial soft tissue loss and facial bone destruction may be extensive in charred remains. Progressive exposure to fire and heat causes the scalp to separate from the calvarium and eventually disintegrate, followed by destruction of the calvarium that begins with the outer table of the skull and ends with the skull base (Figures 6.11 and 6.12). Therefore, even in severely charred remains with substantial thermal destruction of the face and skull, traumatic fractures of the skull base and upper cervical spine may still be evident on postmortem MDCT. Knowledge of this sequence is useful, particularly when cranial or upper cervical trauma is suspected to have occurred prior to death. Traumatic fractures of the skull base and upper cervical spine may be evident on MDCT even when there has been significant thermal destruction of the face and calvarium (Figure 6.16) (Iwase et al. 1998).

The brain as well as all other internal organs shrinks from the heat of a fire. Thermal epidural hematomas are typically biconvex or lentiform in shape. They are characterized by a honeycomb appearance of soft tissue attenuation in the epidural space (Figure 6.9 through Figure 6.11). They extend across a dural sinus, which signifies their location in the epidural space. The honeycomb appearance of a thermal epidural hematoma is due to heat desiccation of the blood. If the epidural is from heat alone, there should be no evidence of traumatic skull fracture. We observed thermal epidural hematomas that contain high-attenuation hemorrhage and are not desiccated on MDCT (Figure 6.13). In these cases, there is no evidence of cerebral trauma on MDCT or autopsy to suggest that the origin of the blood was traumatic. We suspect that the intensity and duration of the fire was such that the epidural hematoma did not desiccate. True traumatic epidural hematomas may have a similar appearance. The distinction can be made when there is evidence of a traumatic skull fracture (Figure 6.14).

Shrinkage and contraction can be observed in all visceral organs exposed to high temperatures. The lungs may show an increase in CT attenuation in addition to a decrease in size (Figure 6.15). The attenuation

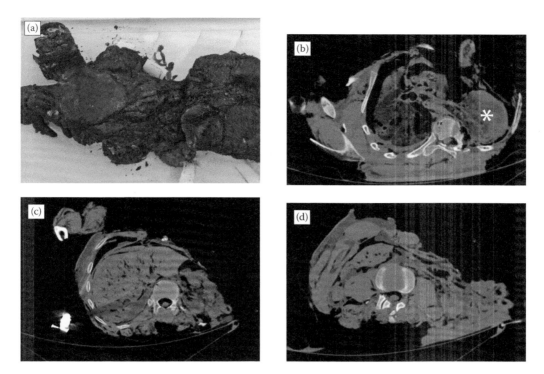

FIGURE 6.15 Torso findings in charred remains. (a) Autopsy photograph shows a severely charred body. (b) Axial MDCT image of the upper chest shows extensive soft tissue loss, thermal loss of the left ribs, and a large defect in the anterior left chest wall. There is thermal shrinkage and increased attenuation of the left lung (asterisk), which is greater than that present in the right lung. A small right hemothorax is present. (c, d) In the upper abdomen and pelvis, there is extensive thermal tissue loss with exposure of the left-sided abdominal contents. Note extensive skin loss throughout.

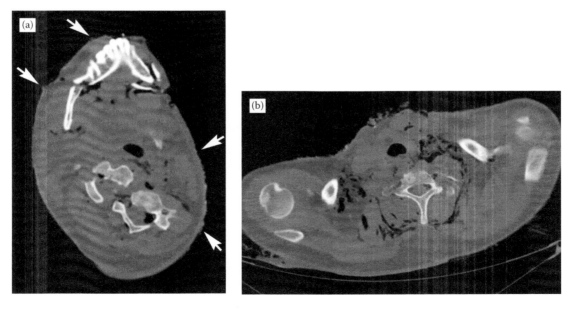

FIGURE 6.16 Antemortem traumatic findings that were the cause of death from blunt trauma before the fire in a victim of an aviation accident. (a) Axial MDCT image of the lower face and neck show a complex cervical spine fracture with transection of the cervical cord. Mandibular fractures are also present. Note the presence of full-thickness burns by the irregular contour of the subcutaneous fat and focal areas of thermal tissue loss (arrows). (b, c) Axial MDCT of the chest shows partial- and full-thickness burns anteriorly and numerous traumatic rib fractures.

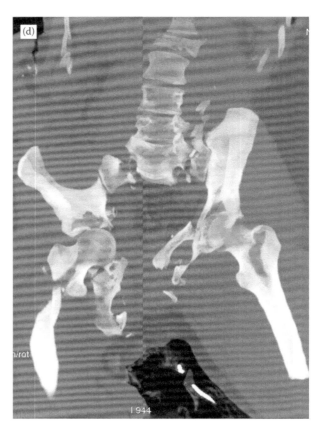

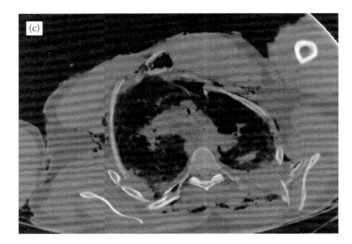

FIGURE 6.16 (*Continued*) Antemortem traumatic findings that were the cause of death from blunt trauma before the fire in a victim of an aviation accident. (b, c) Axial MDCT of the chest shows partial- and full-thickness burns anteriorly and numerous traumatic rib fractures. (d) Coronal MDCT maximal intensity projection image shows a complex fracture of the sacrum, pelvis, and right femur.

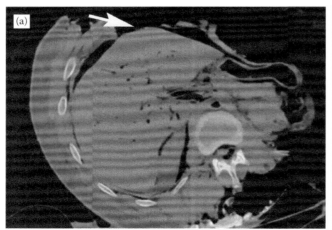

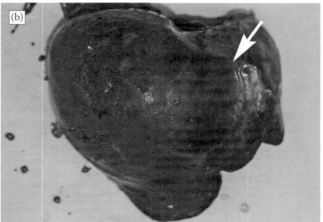

FIGURE 6.17 Asymmetric burns resulting in extensive tissue loss on the left side of the body. (a) Axial MDCT of the upper abdomen shows thermal destruction of the left anterior and lateral body wall with exposure of the visceral organs. Only minimal irregularity is present along the anterior surface of the left lobe of the liver. (b) Autopsy photograph shows charring and blistering of the liver surface (arrow).

change in the lung should not be mistaken for underlying pathologic processes such as pulmonary edema or pneumonia if there is also contraction of the lung parenchyma and other soft tissue findings suggesting thermal injury. Abdominal visceral organs may decrease in size. Surface burns to visceral organs may not be evident on MDCT (Figure 6.17). Although thermal changes in the visceral organs limit the detection of injury and underlying disease, the finding of unsuspected traumatic injury or metallic fragments from a ballistic injury that may have caused or contributed to death facilitates the autopsy of severely burned and charred remains (Figure 6.18).

Uncommonly, MDCT may aid in the identification of charred and fragmented remains through finding distinctive anatomic features, surgical hardware, and localization of tissue suitable for DNA analysis (Harcke et al. 2009). Identification by fingerprinting or dental examination is not possible when charred remains are fragmented and the skull and extremities are no longer present. Deep pelvic structures may be the only remaining recognizable organs (Figure 6.19).

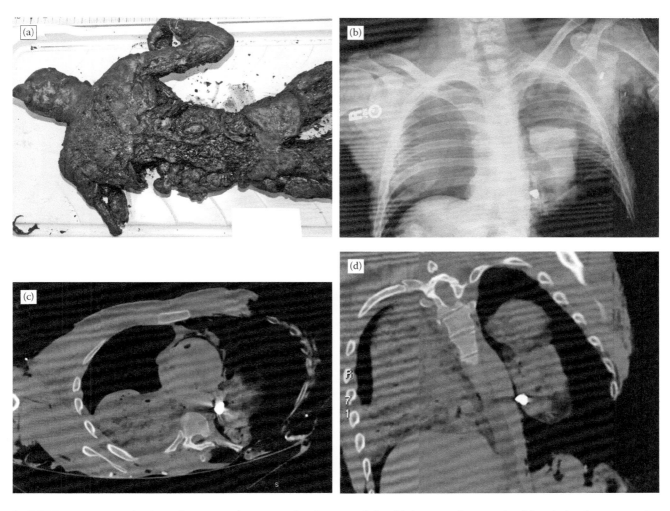

FIGURE 6.18 Severely charred remains of a victim of explosion and fire. (a) Autopsy photograph of the victim shows extensive charring and a pugilistic position of the remaining upper extremity. (b) Anterior-posterior radiograph of the chest shows extensive tissue loss in the left lower chest and upper abdomen and a metallic fragment overlying the medial left lower chest. (c, d): Axial and coronal MDCT show extensive fire-related injury to the thorax and thermal damage to the lungs. An unsuspected metallic fragment from a projectile is present in the medial left lung.

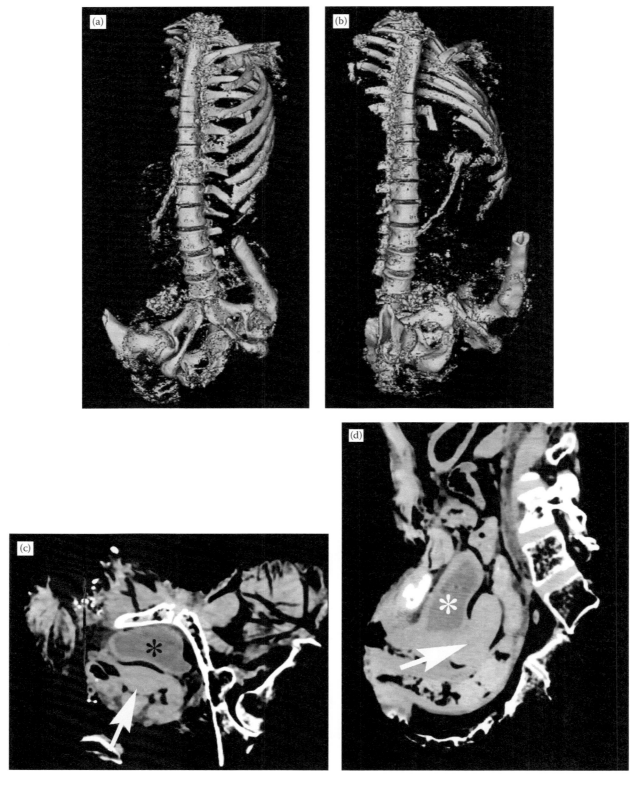

FIGURE 6.19 Victim of a car crash and subsequent fire. (a, b) Three-dimensional MDCT images of a torso show amputation of all extremities and the head with extensive traumatic and thermal destruction to the remaining torso. (c, d) Axial and sagittal MDCT of the lower torso show the uterus is intact (arrows), indicating that this torso is the female passenger of the car. The bladder is marked by an asterisk.

CONCLUSIONS

MDCT can identify the features of postmortem thermal injury in the majority of fire victims. It facilitates autopsy in severely burned and charred remains by accurately displaying traumatic fracture patterns and ballistic fragments that may have occurred antemortem and caused or contributed to death. Postmortem MDCT may serve as a useful preautopsy triage tool in fire-related mass casualty disasters or provide additional anatomic information when the cause of death is rendered by a limited autopsy.

REFERENCES

Administration, U. S. F. 2007. Fire statistics. http://www.usfa.dhs.gov/statistics/ (September 17, 2008).

Bohnert, M., Rost, T., Faller-Marquardt, M., Ropohl, D., and Pollak, S. 1997. Fractures of the base of the skull in charred bodies—post-mortem heat injuries or signs of mechanical traumatisation? *Forensic Sci Int* 87: 55–62.

Bohnert, M., Rost, T., and Pollak, S. 1998. The degree of destruction of human bodies in relation to the duration of the fire. *Forensic Sci Int* 95: 11–21.

Di Maio, V. J. M., and Di Maio, D. J. 2001. *Forensic pathology,* Boca Raton, FL: CRC Press.

Ernst, A., and Zibrak, J. D. 1998. Carbon monoxide poisoning. *N Engl J Med* 339: 1603–1608.

Harcke, H. T., Monaghan, T., Yee, N., and Finelli, L. 2009. Forensic imaging guided recovery of nuclear DNA from the spinal cord. *J Forensic Sci* 54:1123–1126.

Hill, I. R. 1989. Immediate causes of death in fires. *Med Sci Law* 29: 287–292.

Iwase, H., Yamada, Y., Ootani, S. et al. 1998. Evidence for an antemortem injury of a burned head dissected from a burned body. *Forensic Sci Int* 94: 9–14.

Kane, A. B., and Kumar, V. 2005. Environmental and nutritional pathology. In Kumar, V., Abbas, A. K., Fausto, N., Robbins, S. L., and Cotran, R. S. (Eds.) *Robbins and Cotran pathologic basis of disease.* 7th ed. Philadelphia: Elsevier Saunders.

Lawler, W. 1993. Bodies associated with fires. *J Clin Pathol* 46: 886–889.

Levy, A. D., Harcke, H. T., Getz, J. M., and Mallak, C. T. 2009. Multidetector computed tomography findings in deaths with severe burns. *Am J Forensic Med Pathol* 30: 137–141.

Ryan, C. M., Schoenfeld, D. A., Thorpe, W. P. et al. 1998. Objective estimates of the probability of death from burn injuries. *N Engl J Med* 338: 362–366.

Chapter 7

Blast Injury

FORENSIC PRINCIPLES

Deaths from blast injury most often occur in industrial and domestic explosions. In spite of this, the published literature on blast injury is based on experiences from armed conflict. In recent years, acts of terrorism have renewed public, government, and medical interest in blast injury (Langworthy et al. 2004).

Explosion

Explosions are caused by the rapid chemical conversion of solids or liquids into gases, which have a greater volume. The gas generated in an explosion is under high temperature and pressure. The pressure is propagated radially as a mechanical wave into the surrounding area. Kinetic energy is also released from the explosion. The combined effect is termed *brisance*, which is the shattering or crushing effect of the explosion. High kinetic energy or high-velocity explosions generate instantaneous blast waves because they occur in a very condensed time frame that can be measured in milliseconds. High kinetic energy materials often referred to as *high explosive,* generate these types of explosions. They usually require a detonator to start the reaction. A blasting cap is an example of a detonator. Low kinetic energy or low-velocity explosions release their energy slowly by a process called *deflagration*, which has a pushing rather than a shattering blast wave. Gunpowder is an example of a low-velocity explosion (Schwartz et al. 2008).

The blast wave created by an explosion has a pressure front that moves away from the point of the explosion. An initial positive high-pressure peak that immediately decreases and is followed by negative pressure characterizes the pressure front. The negative pressure phase lasts several times longer than the positive pressure phase. In an explosion, the lower-intensity negative pressure can be strong enough to pull in debris from the surrounding area. The negative pressure phase is also known as the *suction phase* (Tsokos 2008).

When the blast front reaches an object, the ambient pressure rises instantly. The highest positive pressure that is

reached is called the *peak overpressure*. The peak overpressure and peak velocity decrease with increasing distance from the center of the explosion. Propagation of overpressure waves is not a simple phenomenon because the surrounding environment affects pressure. A reflected pressure front is created when the pressure wave strikes a solid object and an additive effect can result in a reflected pressure front that is much greater than the incident pressure. The convergence of pressure waves, termed *coupling*, is additive in its destructive power. Reflected pressure fronts and coupling can occur together in closed spaces such as buildings and vehicles as well as in open spaces between buildings such as streets or alleyways. The resulting cumulative overpressure can increase the risk of blast injury. Overpressures of significant magnitude create craters in the ground and nonsurvivable disruption of the body (Sattin et al. 2008).

Underwater explosions produce a shock wave that travels farther than a comparable explosion in air because water is not compressible. The gas created from an underwater explosion forms an expanding bubble that generates a shock wave as the bubble breaks up and rises to the water surface. The blast waves are reflected from the surface of the water and from the bottom of the body of water. Consequently, the blast waves in an underwater explosion are more complex with regard to pressure changes (Bellamy et al. 1991).

Blast Injury Mechanisms

Blast injuries are typically classified as primary, secondary, tertiary, and quaternary. Primary blast injury is secondary to barotrauma; secondary blast injury occurs from penetrating trauma; and tertiary blast injury occurs when the body is propelled through the air and collides with a secondary object, resulting in blunt force injury. Historically, quaternary blast injury included all other blast effects such as thermal, crush, and inhalation injury as well as contamination from chemical, biological, or radiological hazardous materials. More recently, it has been suggested that radiological and biological contamination be considered quintary blast injury (Sattin et al. 2008).

Primary blast injury results from direct exposure of the body to the blast wave. The magnitude of the blast and the distance of the body from the epicenter of the blast determine the amount of peak pressure to which the body is exposed. In addition to distance, the location of the body in a closed or open structure with respect to the blast affects the strength and configuration of the effect of the blast waves on the body. Gas- or air-containing structures within the body are most affected by primary blast injury, because a pressure differential develops between the body surface and internal organs. Pressure within fluid-containing structures remains equal, but air- or gas-containing structures compress and undergo rebound expansion as the blast wave passes. Pressure differentials between blood vessels and air spaces drive fluid into the air spaces. The boundaries of solid soft tissue structures and organs of different density that are adjacent to one another are injured by sheer stresses.

Penetrating trauma characterizes secondary blast injury. It occurs when objects that are accelerated by the energy of the explosion become projectiles and strike the blast victim. Any object can become a projectile in a blast. Commonly, components of the explosive device (e.g., shrapnel intentionally placed in a homemade bomb) and flying debris from the surrounding environment (e.g., collapsed building materials, shattered vehicles, rocks, and vegetation) become injurious projectiles. Penetrating glass fragments are another important component of secondary blast injury, particularly when the explosion occurs in homes or urban environments (Bellamy et al. 1991). The kinetic energy of the projectile, which determines if the victim's skin will be penetrated and the depth of penetration, affects the degree of injury.

The victim's body becomes a projectile in tertiary blast injury. The blast wave pressure differential may throw the body against the ground, equipment, structures, trees, or other stationary objects. Therefore, the location of the victim in the blasts determines the susceptibility to tertiary blast injury. For example, if a person is standing in a doorway or in front of a window where explosive gases vent, that person is vulnerable to tertiary blast injury even if he or she is relatively far from the point of detonation.

Quaternary blast injury includes all other explosion-related injuries, illness, and diseases that are not due to primary, secondary, or tertiary mechanisms. Fire-related injury is one of the most common quaternary injuries. Fires do not originate from the explosion, but from the surrounding environment. Ruptured gas lines, fuel tanks, and damaged electrical circuits are common sources of fire. Fire-related injuries including burns and smoke-related injuries may occur. Other quaternary blast injuries include wound infections from projectile injuries related to embedded glass or debris and bronchospasm or asthma triggered by dust from the explosion. Crush injury and mechanical asphyxiation occurring in structural collapse are also quaternary blast injuries (Kauvar et al. 2008). Mutilating blast injury is defined as traumatic amputation or dissociation of the body from a combination of blast effects.

The probability of blast injury decreases as distance from the epicenter increases. All types of blast injury are usually evident if an individual is very close to the epicenter of the explosion. At distances farther away from the blast, thermal injury becomes less likely. Projectile injuries continue to occur and account for the majority of injuries.

AUTOPSY FINDINGS

The autopsy findings of blast injury include the spectrum of injuries related to the four mechanisms of blast injury. Victims of blast injury may be intact and show no or minor findings on the skin surface, have one or more amputations, or have total body disruption. In the latter instance, dissociated body parts are received. These are usually mixed with debris and may contain parts from several different individuals. Amputation results in a lesser degree of body disruption and fragmentation. Amputations include loss of digits, extremities, and the head (Figure 7.1). Fully intact remains may show evidence of fatal injury or injury that was initially survivable. In all cases, the remains may show evidence of projectile injury, blunt force injury, or burns (Figure 7.2). The task of the forensic pathologist is to examine the component injuries with regard to lethality, types of injury, and recovery of evidence to assist in the investigation.

Primary blast injury affects the auditory, respiratory, and circulatory systems, and hollow viscera. The organs become distorted from pressure, which creates tissue stress. When tissue stress exceeds tissue strength, injury occurs. Air-containing organs rely on natural pressure-equilibrating mechanisms to maintain structure and

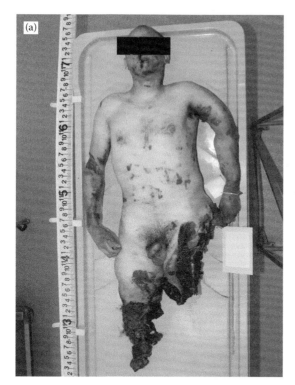

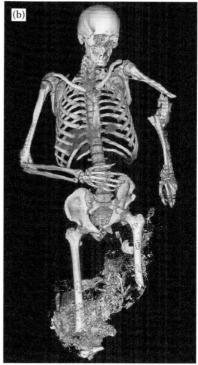

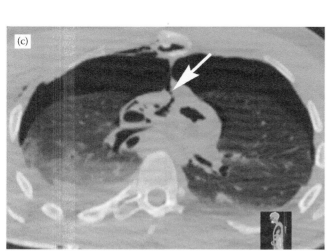

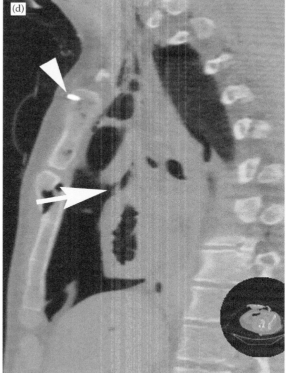

FIGURE 7.1   Blast injury with traumatic amputation. (a, b) Autopsy photograph and three-dimensional MDCT show bilateral above-the-knee amputations with associated fractures of the pelvis and left arm. (c, d) Axial and sagittal MDCT of the thorax show bilateral pneumothorax and hemothorax. There is a laceration of the anterior wall of the heart (arrows) and fracture of the sternum. The metal tip of an intraosseous intravenous apparatus is seen in the manubrium of the sternum (arrowhead). Intravascular air is secondary to decomposition.

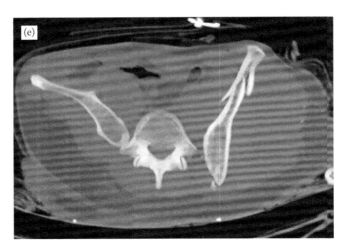

FIGURE 7.1 (*Continued*) Blast injury with traumatic amputation. (e) Axial MDCT of the pelvis shows fractures of the sacroiliac joints and the left iliac bone.

functionality. These are overcome by the rapid change in external pressure in an explosion (Bellamy et al. 1991).

In the auditory system, the tympanic membrane is the most susceptible to blast injury. It may rupture at relatively low pressures. Consequently, a tympanic membrane rupture is almost always found at pressure levels that are high enough to injure the lungs and hollow gastrointestinal viscera. Ossicular fracture and cochlear damage may also occur. The ear closest to the blast usually has the most severe damage (Tsokos 2008).

Pulmonary edema, contusion, laceration, and hemorrhage occur in blast injury to the respiratory system. The lungs are especially vulnerable to blast injury because of their structure and location. The relatively rigid rib cage can be displaced into the lung during the blast causing brief compression of the lungs. Contusions may result. These are located on the pleural surface and may show distinctive impressions that are often referred to as *rib markings*. The source of the markings is actually the intercostal muscle. Pale bands represent tissue and vascular structures compressed by the ribs (Figure 7.3). The term *intercostal markings* is more appropriate for this finding (Bellamy et al. 1991).

Pulmonary parenchymal hemorrhage is related to rupture of alveoli septae and tearing of alveolar capillaries with subsequent bleeding into the alveoli. Histologically, pulmonary hemorrhage is characterized by blood-filled alveoli. Alveolar septal tears are often difficult to identify. At autopsy, pulmonary hemorrhage is manifest as oozing of blood from the cut surface of the lung as it is sectioned (Figure 7.3) (Bellamy et al. 1991). Alveolovenous fistulas may occur when alveolar wall tears involve an intralobular venule. Direct communication between air-filled alveoli and the circulatory system may result in air emboli. It has been stated that air emboli may be the primary cause of immediate death in blast victims and the cause of most primary blast injury deaths rather than alveolar hemorrhage (Bellamy et al. 1991). Pulmonary emphysema may occur when alveolar septa tears coalesce and create larger air collections. Pneumothorax and hemothorax occur from pleural rupture (Figure 7.1). Primary blast injury may also cause ulceration and submucosal hemorrhage in the larger bronchi and upper airways when their epithelium is stripped from the underlying basal lamina (Bellamy et al. 1991).

The circulatory system is affected directly by blunt force injury or indirectly by air emboli originating in the lung. Air emboli can occlude coronary blood arteries and cause cardiac arrest or travel to the cerebral arteries to produce neurological injury. Most investigators believe that clinically significant air emboli occur within the first 10 minutes of the blast (Bellamy et al. 1991). Pathologic confirmation of air emboli is difficult. The diagnosis should be considered prior to autopsy. The finding of air isolated to the heart or cerebral vasculature on radiographs or multidetector computed tomography (MDCT) suggests the diagnosis. Air emboli are difficult to document at autopsy because they require special techniques of submerging the heart in water or

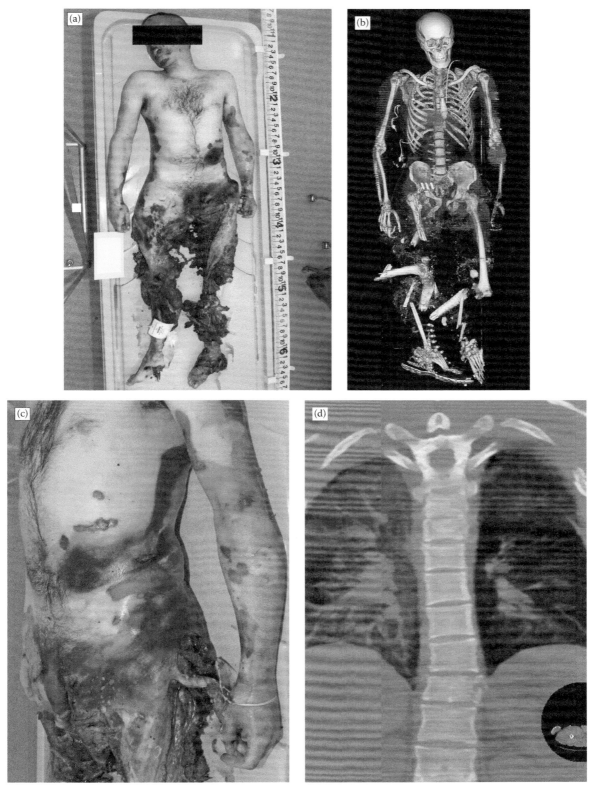

FIGURE 7.2 Secondary, tertiary, and quaternary blast injury. (a, b): Autopsy photograph and three-dimensional MDCT show soft tissue disruption of the lower extremities with multiple comminuted fractures. The left arm is fractured. (c) Autopsy photograph of the left side of the torso shows partial- and full-thickness burns over the lower abdomen, pelvis, thighs, and back. (d, e) Coronal and sagittal MDCT of the thorax show multiple thoracic vertebral fractures and fracture of the sternum. There is a left pneumothorax.

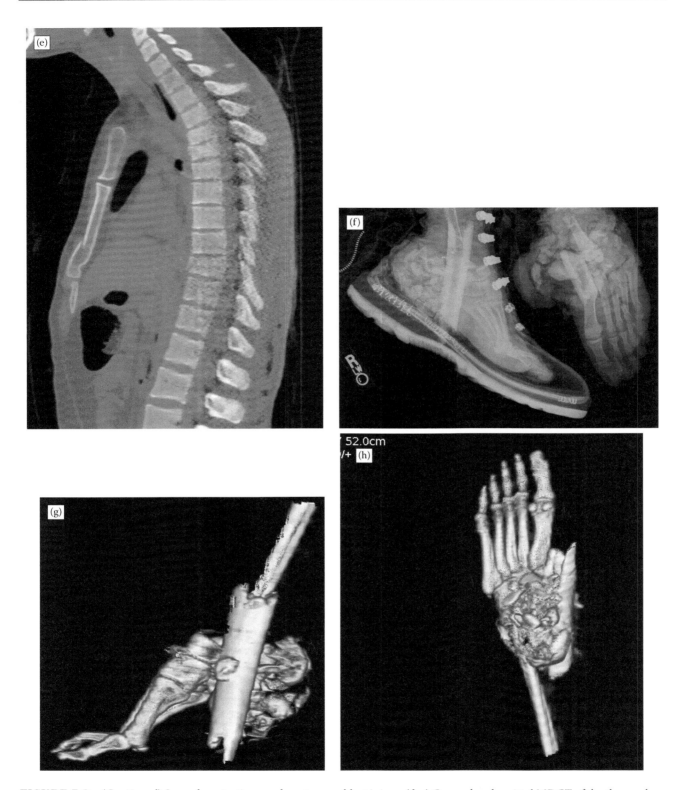

FIGURE 7.2 (*Continued*) Secondary, tertiary, and quaternary blast injury. (d, e) Coronal and sagittal MDCT of the thorax show multiple thoracic vertebral fractures and fracture of the sternum. There is a left pneumothorax. (f) Digital radiograph of the feet shows complex hindfoot, midfoot, and ankle fractures bilaterally. (g, h, i, j) Three-dimensional MDCT and autopsy photographs show cranial displacement of the foot and "spearing" of the foot by the tibia, indicating that the blast came from underneath the feet.

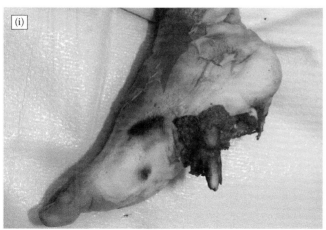
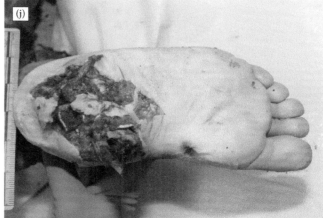

FIGURE 7.2 (*Continued*) Secondary, tertiary, and quaternary blast injury. (g, h, i, j) Three-dimensional MDCT and autopsy photographs show cranial displacement of the foot and "spearing" of the foot by the tibia, indicating that the blast came from underneath the feet.

an aspirometer. Even with these techniques, postmortem gas production as part of the decomposition process may make these findings difficult to interpret. Cardiac hemorrhage from primary blast injury may occur when displacement of the mediastinum causes cardiac laceration or contusion. Cardiac hemorrhage is rarely seen without accompanying pulmonary hemorrhage (Figure 7.1) (Bellamy et al. 1991).

Injuries to the gastrointestinal tract are more likely to occur when blast waves are propagated through water than through air (Bellamy et al. 1991). Gastrointestinal injuries include intramural hemorrhage, laceration, and rupture. The distal small intestine and large intestine, particularly the cecum, are most vulnerable because they commonly contain the most gas. Bowel perforation may occur immediately or be delayed many days after blast exposure due to injury and subsequent necrosis of the bowel wall.

Blast-related neurotrauma is currently receiving intense investigation because the effects on survivors are poorly understood. Emerging evidence suggests that blast pressure transmitted transcalvarially may affect brain function. Overpressure, however, is only one potential etiology for traumatic brain injury. Electromagnetic energy, acoustics, and other etiologies have been implicated (Ling et al. 2008).

Projectiles and debris that cause secondary blast injury may inflict blunt and penetrating trauma on blast victims depending on the size, configuration, and velocity of the projectile. A classic triad of contusions, abrasions, and lacerations characterizes secondary blast injury. Internally, projectile fragments follow ballistic principles, and the degree of tissue damage is dependent on the kinetic energy released and the path of the projectile (Figure 7.4).

The injuries from tertiary blast trauma depend on the velocity imparted to the body by the blast wave. Survivable injury occurs when the body impacts an object at less than 10 feet per second. Velocities over 20 feet per second are the thresholds for lethal injury when body impact is against a hard, flat surface. Body impacts caused by velocities over 30 feet per second are reported to be 100% lethal (Schwartz et al. 2008). The skull and brain are more vulnerable to injury and, consequently, have lower lethality thresholds. At autopsy, it may not be possible to determine if contusions and abrasions on the external surface are from secondary or tertiary blast injury. Internally, fractures and organ damage may occur as a result of projectile injury or blunt force injury (Figure 7.5). Accordingly, most forensic pathologists usually describe these findings as *blunt force injury,* and the exact mechanism in many cases can be difficult to determine (Figure 7.6).

Quaternary blast injury should be assessed in conjunction with other injuries to establish its role in the cause of death. The findings of burns related to a blast are the same as burns from other causes of fire, which are discussed in Chapter 6.

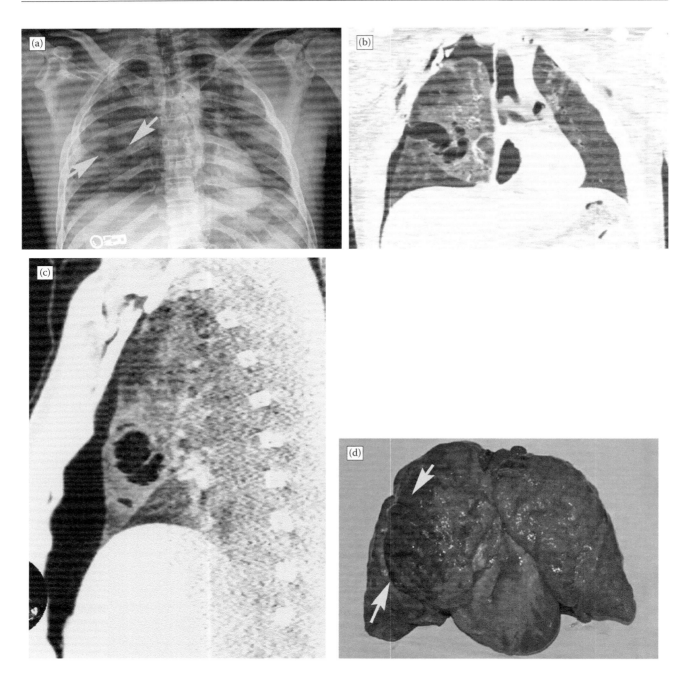

FIGURE 7.3   Blast lung. (a) Postmortem chest radiograph shows fractures of the right scapula and first rib and a right pneumothorax with shift of the cardiac and mediastinal structures to the left. There is a focal lucent air collection in the right lower lung (arrows). (b, c) Coronal and sagittal MDCT images show that the right lower lung lucency observed on the chest radiograph represents an air-filled cavity from lung laceration. There is increased attenuation surrounding the cavity which represents hemorrhage. (d) Photograph of the lungs and mediastinum removed at autopsy shows striped contusions (arrows) on the surface of the right middle lobe secondary to intercostal impressions at the time of the blast.

RADIOLOGIC PRINCIPLES

Goals of Imaging

Whole-body imaging is especially well suited for victims of blast injury. The spectrum of blast injury includes effects that have the potential to focally or diffusely damage tissue from head to toe (Figures 7.1 and 7.2). Radiography is useful to obtain a quick overview of the extent of gross disruption, such as amputation, as well as the distribution of metallic fragments that may be present. In catastrophic blast injury,

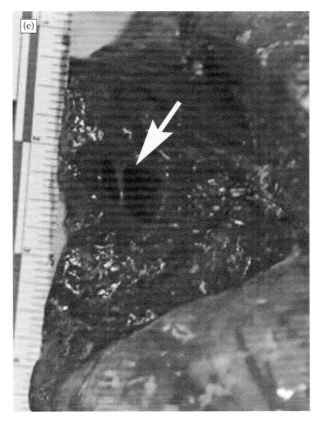

FIGURE 7.3 (*Continued*) Blast lung. (e) Photograph of the cut surface of the right lung shows the lung cavity (arrow) and surrounding hemorrhage.

lethal injuries are easily evident. Cross-sectional imaging with MDCT is most useful as an adjunct to autopsy when the cause of death is less obvious, particularly if there is minimal evidence of external injury (Figure 7.7).

Imaging Findings

The extent and severity of blast injury depend on the victim's proximity to the blast and whether or not the victim was sheltered in a building or vehicle, in the open, or in water. Radiographs frequently show natural and man-made debris on the skin surface or embedded within the body of a blast victim. The forms, shapes, and radiographic densities of debris are variable. MDCT affords three-dimensional imaging, which is helpful in differentiating superficial and external debris from material within the body (Figure 7.8). As mentioned in other chapters, radiography is better than MDCT for recognizing the shape of metallic objects and fragments (Figure 7.7). Natural material fragments, such as rocks, will often penetrate the body if they have sufficient velocity. On MDCT, rocks are typically homogeneous in attenuation and may be less dense than bone.

Primary blast injury occurs most commonly in the lungs and abdomen where gas- and air-containing organs are found. In principle, primary and tertiary blast injuries are separate, but it is neither possible nor practical to differentiate them on imaging studies. Pulmonary laceration and hemorrhage may occur in both primary and tertiary blast injury. Pulmonary hemorrhage causes increased attenuation in the lung parenchyma on MDCT. Hemorrhage may be focal or diffuse. Pulmonary lacerations typically cause a more focal or linear area of increased attenuation within the lung which represents a focal hematoma. Occasionally, air may be visualized within a laceration (Figure 7.3). Adjacent soft tissue structures and organs may also lacerate and hemorrhage as a result of barotrauma in primary blast injury or from blunt trauma in tertiary blast injury. Rib fractures are common in blast injury and almost always indicate blunt force injury. In our experience, rib fractures are

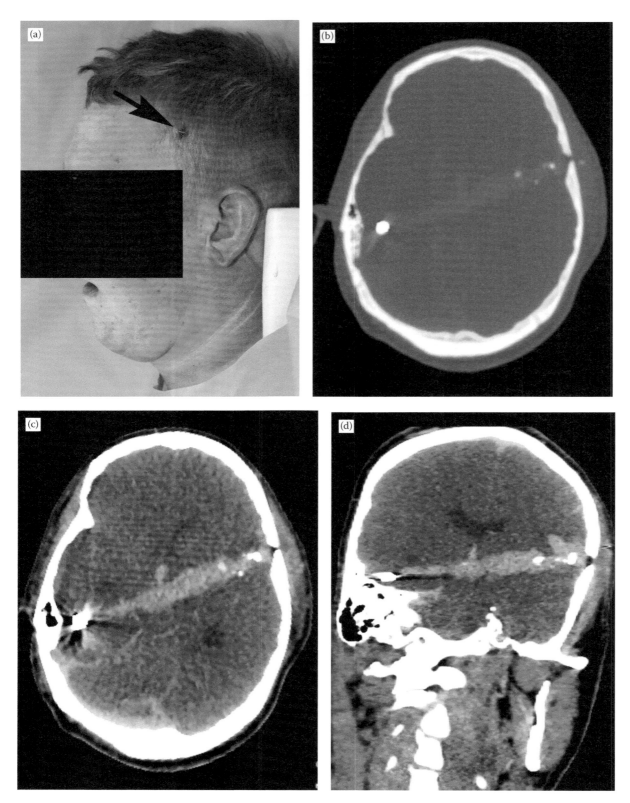

FIGURE 7.4   Secondary blast injury, a penetrating metal fragment. (a) Autopsy photograph shows left temporal entry wound (arrow). (b) Oblique axial MDCT in plane of the wound track shows internal beveling and bone fragments at skull entry. The metal fragment is in the right posterior temporal brain. (c, d) Oblique axial and coronal MDCT in plane of wound track show hemorrhage that documents the course of the projectile: left to right, anterior to posterior, and downward.

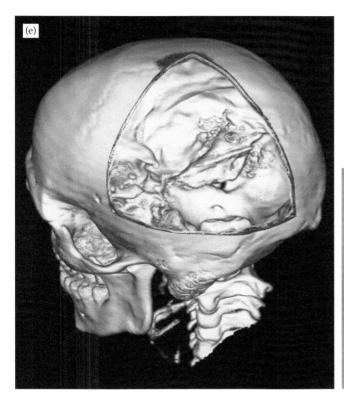

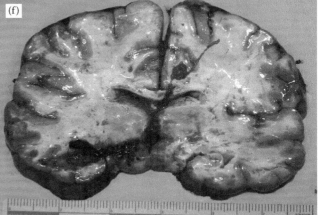

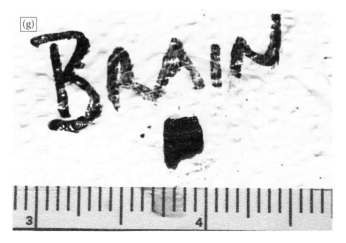

FIGURE 7.4 (*Continued*) Secondary blast injury, a penetrating metal fragment. (e) Three-dimensional MDCT with removal of a calvarial flap to expose the fragment location (fragment shown in red) within the cranial cavity. The entry site is in the left temporal bone. (f) Autopsy photograph of the cut surface of the brain displayed to match computed tomography. Orientation shows final temporal path. (g) Recovered fragment.

invariably present in conjunction with pulmonary hemorrhage and are also often associated with laceration of the pulmonary parenchyma. Therefore, the findings of pure primary blast lung injury are rare in our experience. Even when there is no evidence of rib fracture, the occurrence of blunt force injury elsewhere may suggest that pulmonary findings may be related to tertiary blast injury (Figure 7.1 and Figure 7.7).

MDCT imaging of the brain after blast injury readily shows penetrating head trauma involving fracture and disruption of the calvarium and extra-axial hemorrhage from dural

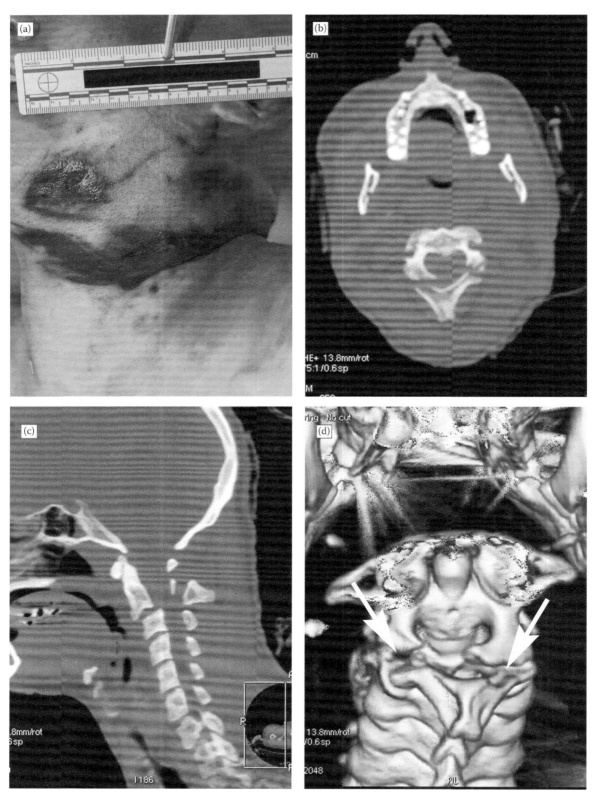

FIGURE 7.5   Tertiary blast injury to the cervical spine. (a) Autopsy photograph shows contusions of the cervical region and contusions and lacerations on the mandible and below the left ear. (b) Axial MDCT shows fractures of the right and left lamina of C2. (c) Sagittal maximum intensity projection shows anterior subluxation of C2 on C3. (d) Three-dimensional MDCT with computer subtraction of the skull and posterior arch of C1 to show the posterior C2 fractures (arrows).

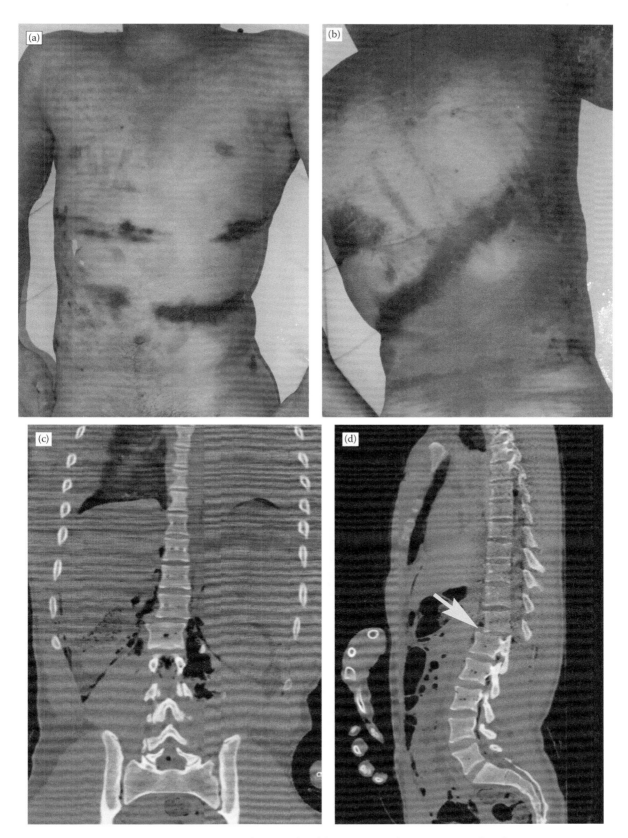

FIGURE 7.6   Tertiary blast injury. (a, b) Autopsy photographs of the anterior and posterior torso show linear contusions consistent with blunt force injury. (c, d) Coronal and sagittal MDCT show retrolisthesis of T12 on L1 (arrow in d).

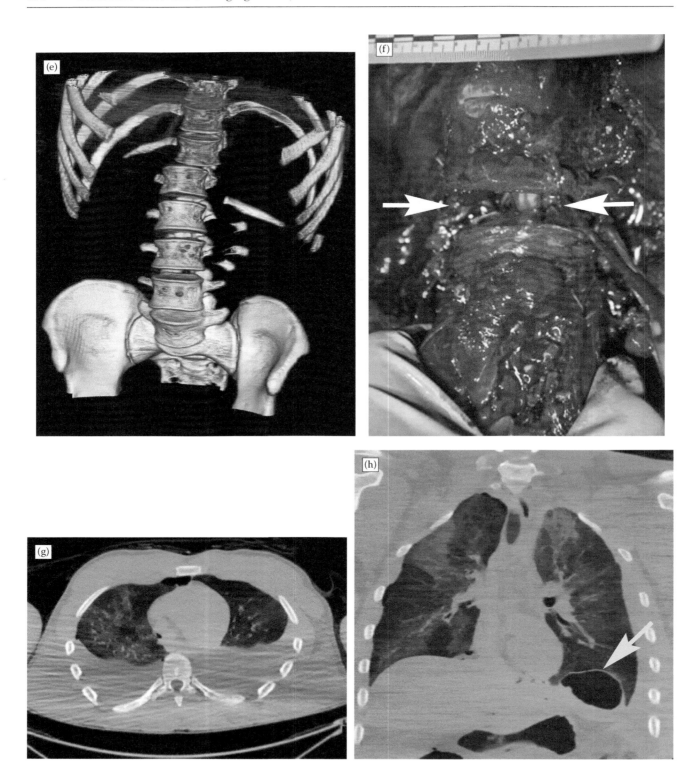

FIGURE 7.6 (*Continued*) Tertiary blast injury. (e) Three-dimensional MDCT shows fractures of the left 12th rib and the transverse processes of L1 through L3. The dislocation of the T12/L1 disk level is also evident on this image. (f) Autopsy photograph shows separation of the spine at T12/L1 disk space (arrows). (g, h, i) Axial, coronal, and sagittal MDCT of the chest and upper abdomen show traumatic herniation of the stomach (arrows) into the lower chest from laceration of the left hemidiaphragm. Patchy, multifocal areas of increased attenuation are present in both lungs from pulmonary hemorrhage. There is a small right and left pneumothorax present and bilateral hemothoraxes.

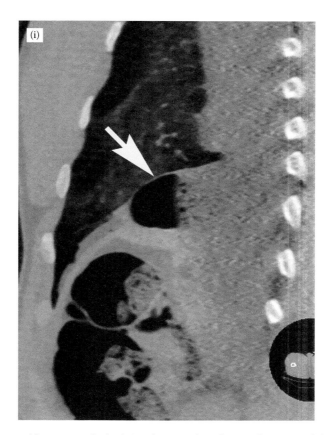

FIGURE 7.6 (*Continued*) Tertiary blast injury. (g, h, i) Axial, coronal, and sagittal MDCT of the chest and upper abdomen show traumatic herniation of the stomach (arrows) into the lower chest from laceration of the left hemidiaphragm. Patchy, multifocal areas of increased attenuation are present in both lungs from pulmonary hemorrhage. There is a small right and left pneumothorax present and bilateral hemothoraxes.

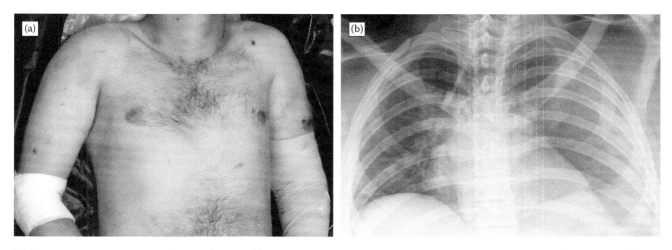

FIGURE 7.7 Primary, secondary, and tertiary blast injury. (a) Autopsy photograph of the torso shows no significant external bruising or laceration. Both elbows are bandaged. (b) Chest radiograph shows mild generalized increased density overlying the left lung and shift of the mediastinal and cardiac structures to the right. The heart is enlarged.

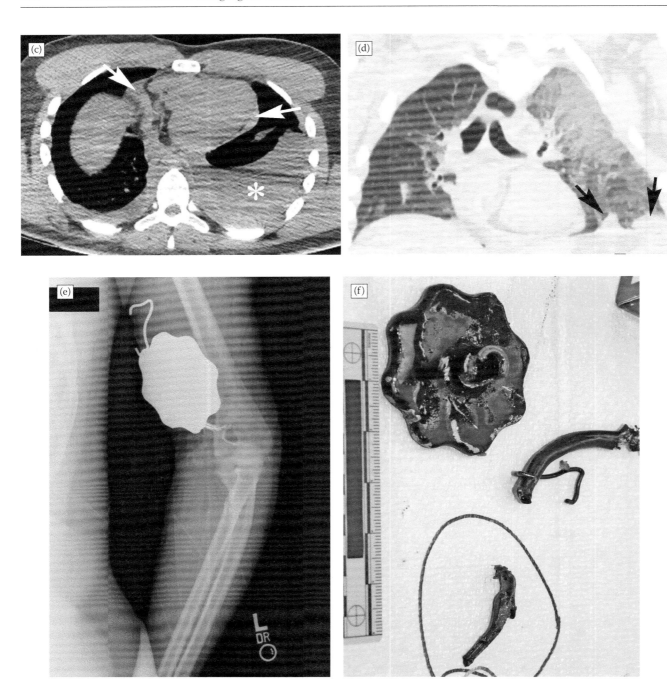

FIGURE 7.7 (*Continued*) Primary, secondary, and tertiary blast injury. (c, d) Axial and coronal MDCT images of the chest show a hemopericardium (arrows in c) and a large left hemothorax (asterisk in c). Lung windows (image d) show multiple areas of hemorrhage in both lungs and focal left lower lobe hematomas from laceration (arrows in d). (e) Radiograph of the left arm shows embedded metallic objects. (f) Photograph of the recovered metallic objects shows the objects are a valve handle and wire.

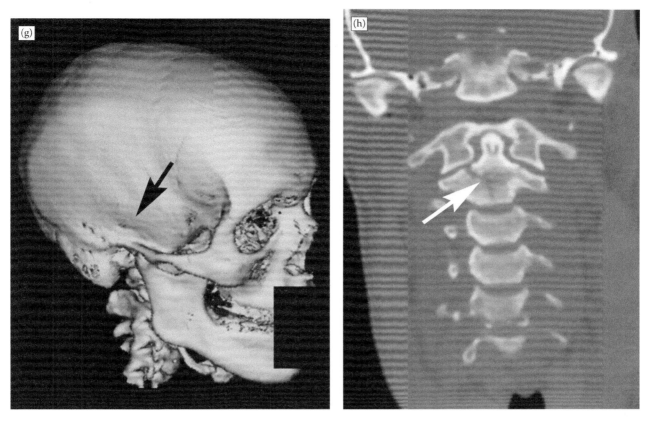

FIGURE 7.7 (*Continued*) Primary, secondary, and tertiary blast injury. (g, h) Three-dimensional and coronal MDCT show the head and neck findings of a linear right temporal skull fracture (arrow in g) and fracture of C2 at the base of the odontoid process (arrow in h).

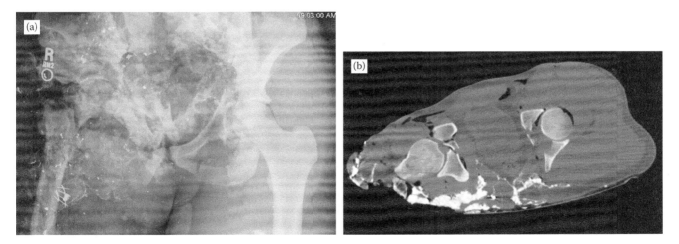

FIGURE 7.8 Tertiary and quaternary blast injury. (a) Frontal radiograph of the pelvis shows complex, comminuted fractures of the pelvis and proximal femurs. There are a variety of high densities overlying the right hemipelvis and soft tissues. (b) Axial MDCT shows that the high density identified on the radiograph represents superficial metallic spalling, which is associated with the blast and fire. Thermal tissue loss is present in the posterior soft tissues of the right pelvis. Bilateral acetabular fractures are also present.

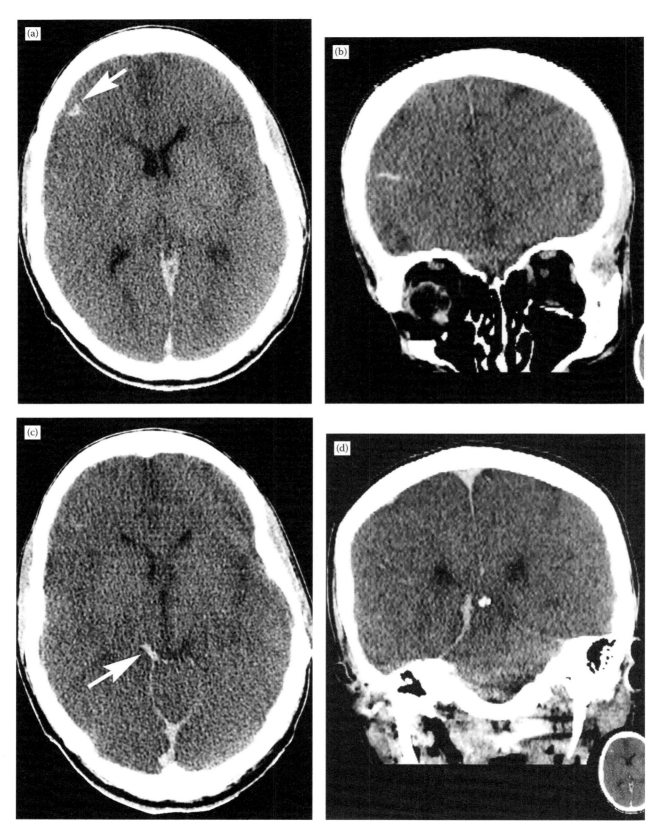

FIGURE 7.9   Closed head injury from blast. (a, b): Axial and coronal MDCT show a focal contusion in the right frontal lobe (arrow). (c, d) Axial and coronal MDCT show subarachnoid hemorrhage in the right interpeduncular cistern (arrow).

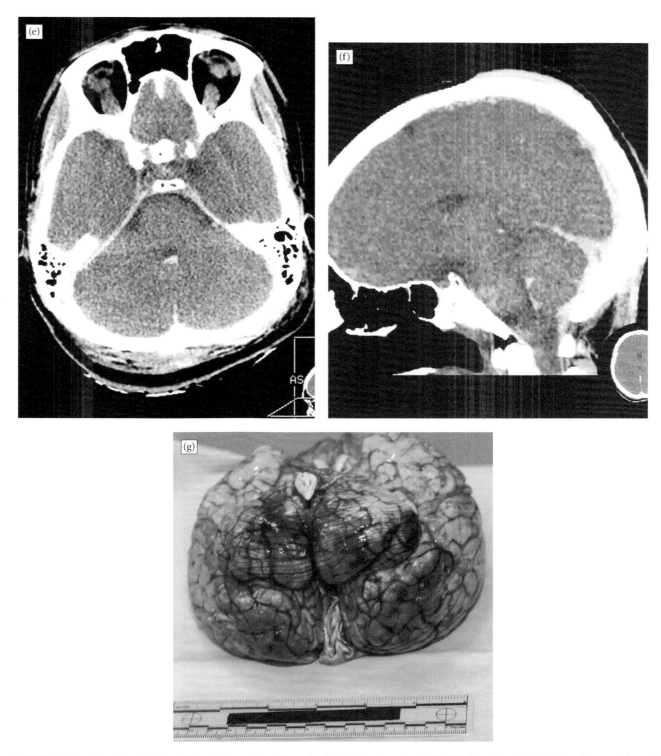

FIGURE 7.9 (*Continued*) Closed head injury from blast. (e, f) Axial and sagittal MDCT show blood in the fourth ventricle and cerebellomedullary cistern. (g) Autopsy photograph shows occipital and cerebellar subarachnoid hemorrhage.

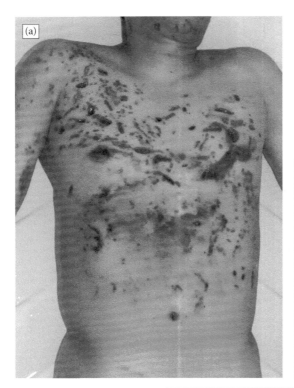

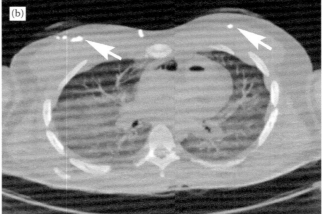

FIGURE 7.10 Secondary blast injury with embedded glass. (a) Autopsy photograph of the torso shows multiple lacerations and abrasions. (b) Axial MDCT shows medium attenuation fragments (arrows) in the anterior chest wall. There is air space consolidation within the right lung and bilateral posterior congestion consistent with decomposition or blast. (c) Photograph of the glass shards recovered from the anterior chest wall.

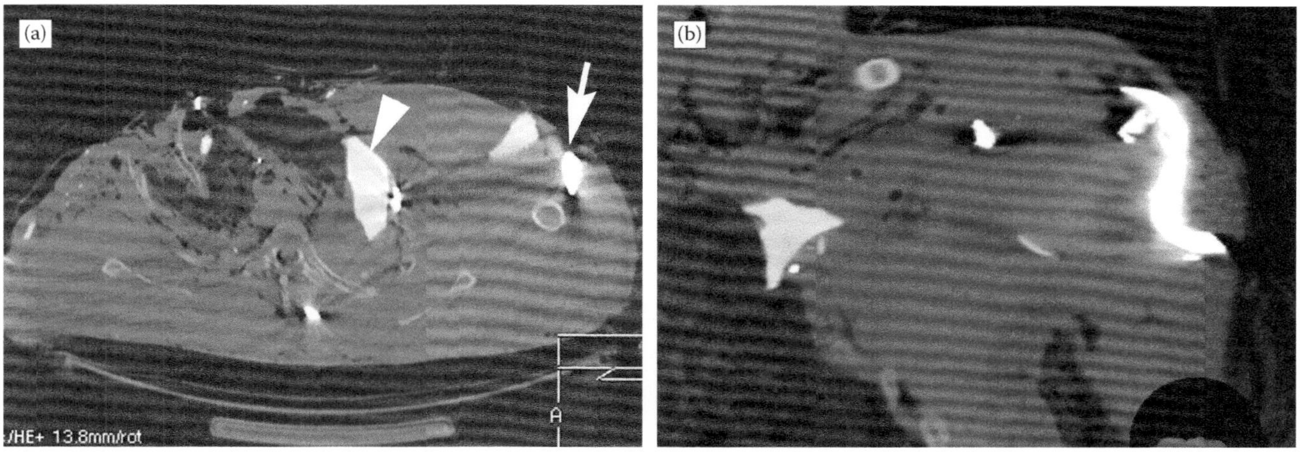

FIGURE 7.11 Secondary blast injury with various metallic projectiles. (a) Axial and (b) coronal MDCT of the chest show multiple large metallic fragments. Some show very high attenuation and streak artifact (left shoulder) (arrow), and others have high attenuation without significant streak artifact (chest wall) (arrowhead). Metallurgic evaluation showed that the former were composed of steel, and the latter were an alloy of aluminum. The image also shows disruption of the chest with rib fractures, hemothoraxes, and mediastinal soft tissue disruption.

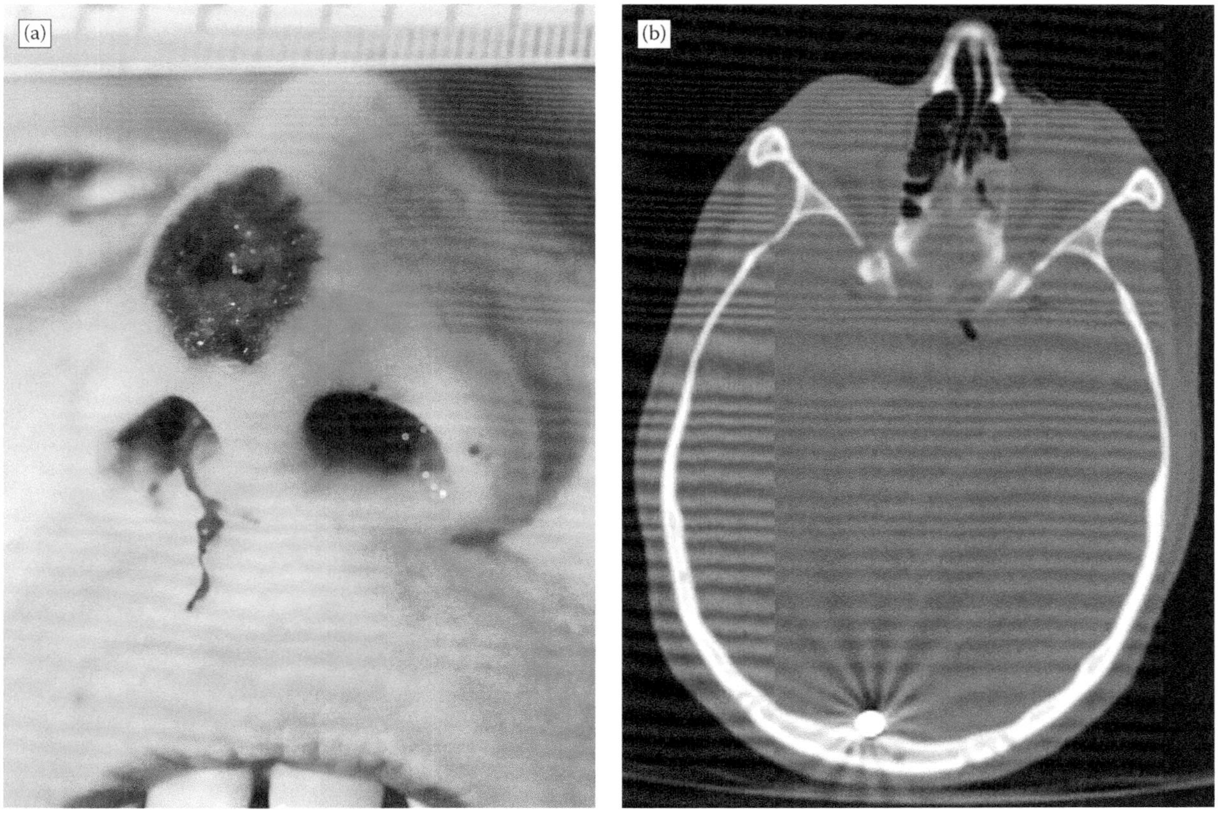

FIGURE 7.12 Secondary blast injuries in a suicide-bombing victim. Steel ball bearings incorporated into the explosive device caused projectile injury. (a) Autopsy photograph shows a projectile entry wound in the nose. (b) Axial MDCT shows a rounded metallic projectile in the posterior brain against the inner table of the occiput.

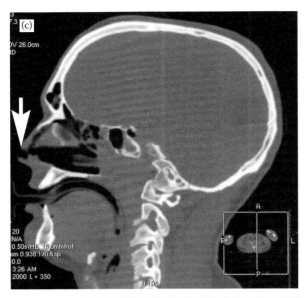

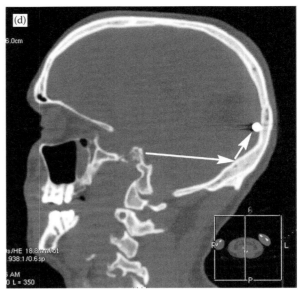

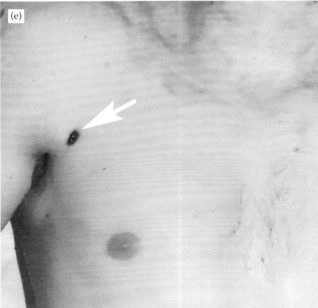

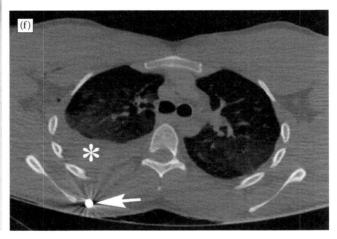

FIGURE 7.12 (*Continued*) Secondary blast injuries in a suicide-bombing victim. Steel ball bearings incorporated into the explosive device caused projectile injury. (c, d) Sagittal MDCT images in the wound path show the nasal entry wound (arrow in c) and that the projectile passed into the posterior fossa through the inferior portion of the petrous ridge. The fractures of the petrous ridge identify it as an intermediate target. The final location of the projectile is superior to the wound path, indicating that its path was altered by the impact on the occipital bone (arrows indicate probable projectile course). (e, f) Autopsy photograph of the right anterior chest and axial MDCT of the chest show that another projectile entered the right chest anterior to the axilla (arrow in e). It passed through the right hemithorax creating a hemopneumothorax (asterisk in f) and rib fractures (not shown in this image) before coming to rest medial to the right scapula (arrow in f).

tears. Closed head injury with parenchymal hemorrhage, increased intracranial pressure, and contusion can be recognized in the early postmortem period before decomposition obscures the findings (Figure 7.9). Blast overpressure neurotrauma has not been characterized on postmortem MDCT. Even when subtle areas of hemorrhage are found

in the absence of fracture, it is generally impossible to separate primary from tertiary mechanisms of injury.

Evaluation of secondary blast injury from projectiles can be approached in the same manner that gunshot injuries are studied, because they both represent ballistic injury

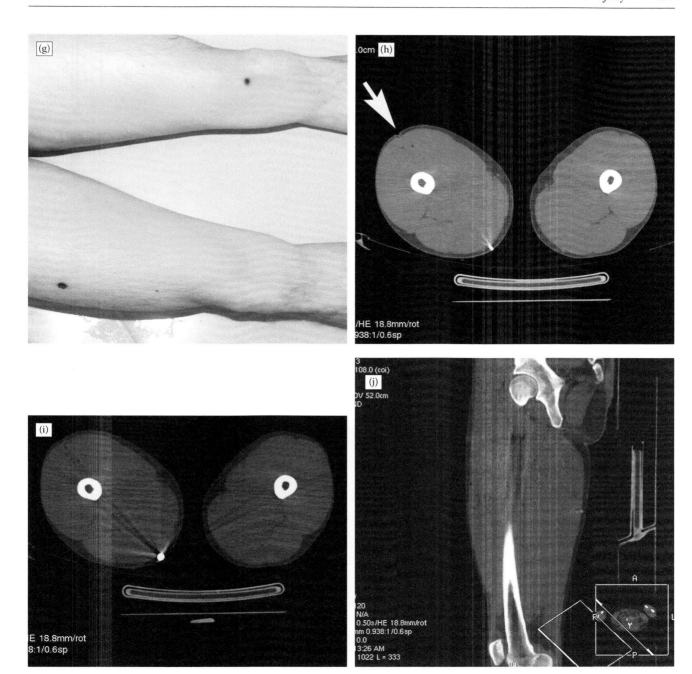

FIGURE 7.12 (*Continued*) Secondary blast injuries in a suicide-bombing victim. Steel ball bearings incorporated into the explosive device caused projectile injury. (g) Autopsy photograph shows wound is on the right thigh and above the left knee. (h, i, j) Axial and oblique sagittal MDCT of the right thigh show the anterior entry wound (arrow) and ball bearing at the posterior skin surface. Some scattered air collections in the soft tissue reflect the wound track that missed the femur.

(Figure 7.4). Both penetrating and perforating wounds occur in blasts, and they will often be present concurrently. All sizes, shapes, and types of materials can be found as projectiles. The composition of some materials can be more easily recognized on radiographs compared to MDCT, because radiographs have superior edge detail

(Figure 7.7). On MDCT, the attenuation of projectile fragments is also variable because both natural and man-made materials can be projectiles. As discussed earlier, glass is a common secondary projectile. It can be identified on MDCT because it has an attenuation that is similar to cortical bone but lacks streak artifact that is commonly

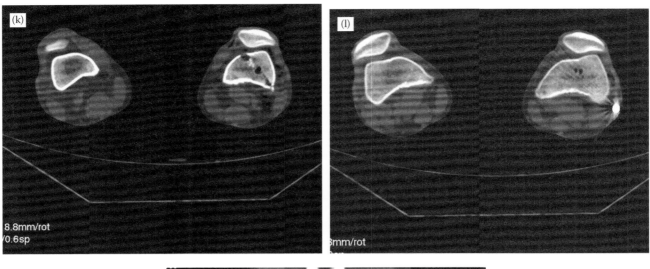

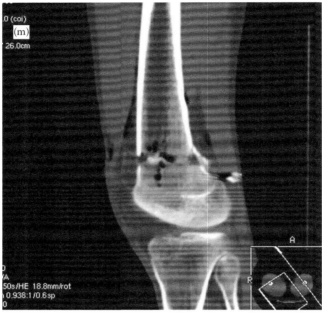

FIGURE 7.12 (*Continued*) Secondary blast injuries in a suicide-bombing victim. Steel ball bearings incorporated into the explosive device caused projectile injury. (k, l, m) Axial and oblique sagittal MDCT of the left knee show a ball bearing passed through the distal femur (anterior to posterior, right to left, and downward) and is lodged in the posterior lateral soft tissue.

observed in metallic fragments (Figure 7.10). It is not possible to be specific as to the composition of metal fragments or glass. For example, steel and copper have similar attenuation values and cannot be differentiated readily on MDCT. Furthermore, many materials are metallic alloys. We found that it is useful to comment on the relative differences in fragment attenuation on MDCT when present so that the medical examiner can be sure to retrieve two different types of fragments from the body (Figure 7.11). Projectiles may be intentionally incorporated into

explosive devices that are military ordinance, terrorist bombs, or homemade bombs. Suicide bombs used in terrorist activity often contain multiple metallic fragments such as ball bearings or nails because they increase the lethality of the blast (Figure 7.12) (Aharonson-Daniel et al. 2008). It may be possible to estimate the location and direction of the blast by studying the pattern of projectile distribution and wound tracks in the victim. This information can prove helpful to forensic investigators and law enforcement.

Tertiary blast injury is essentially blunt force injury. MDCT is most useful for detecting traumatic injuries in areas that are not readily accessed in the autopsy. Fractures of the spine and pelvis are of foremost interest because of their association with lethal injury (Figure 7.3 through Figure 7.6). Analysis of fracture patterns may also allow the direction of the blast to be determined in many cases. For example, lower extremity fractures caused by explosive forces from below show features of axial load injuries of the ankle and foot, such as the long bones of the lower extremity forced through the mid- and hindfoot (Figure 7.2).

Quaternary blast injury is found in conjunction with injuries from the other mechanisms. Burns are one of the most common quaternary injuries in blast victims. The severity of primary, secondary, and tertiary blast injury should always be carefully considered when determining the cause of death in a blast victim who has significant burns, because fire-related injury may have occurred after death (Figure 7.2). Furthermore, it is not uncommon to observe partial burns in blast victims or spalling of metallic particles that coat the surface of the burned area (Figure 7.8).

The destructive nature of explosions often results in mass casualty events where there are numerous victims of blast injury. The finding of fragmented and commingled remains at the scene makes victim identification more difficult.

CONCLUSIONS

The application of MDCT to the postmortem examination of blast injury provides medical examiners with information about bone and soft tissue damage in areas that are difficult to visualize at the standard forensic autopsy. Primary, secondary, tertiary, and quaternary blast injuries frequently occur at the same time. Consequently, the forensic and imaging principles applied to the assessment of ballistic, blunt force, and thermal injury are utilized in blast cases. MDCT greatly enhances the accuracy and speed of documenting the details of fractures resulting from secondary and tertiary blast injury of the spine and pelvis. Postmortem

MDCT provides accurate assessment of blast injury in air-containing organs such as the lungs. However, the assessment of solid visceral organs such as the liver, spleen, and kidneys is limited. Major laceration and pulpification are readily identified, but assessment of minor lacerations and vascular damage is limited. Total-body imaging with MDCT provides an excellent overview of the extent and severity of blast injury and quickly provides a guide to injuries that are most likely to be lethal. MDCT is potentially very useful in the management of explosions that create mass casualties. There is great potential for the use of MDCT in mass casualty incidents from explosions.

REFERENCES

Aharonson-Daniel, L., Almogy, G., Bahouth, H. et al. 2008. Mass casualty events—suicide bombing: the Israeli perspective. In Elsayed, N. M., and Atkins, J. L. (Eds.), *Explosion and blast injuries: effects of explosion and blast from military operations, industrial accidents, and acts of terrorism.* Boston: Elsevier Academic Press.

Bellamy, R. F., Zajtchuk, R., and Buescher, T. M. 1991. *Conventional warfare: ballistic, blast, and burn injuries,* Washington, DC: Walter Reed Army Institute of Research.

Kauvar, D. S., Dubick, M. A., Blackbourne, L. H., and Wolf, S. E. 2008. Quaternary blast injury: burns. In Elsayed, N. M., and Atkins, J. L. (Eds.), *Explosion and blast injuries: effects of explosion and blast from military operations, industrial accidents, and acts of terrorism.* Boston: Elsevier Academic Press.

Langworthy, M. J., Sabra, J., and Gould, M. 2004. Terrorism and blast phenomena: lessons learned from the attack on the USS Cole (DDG67). *Clin Orthop Relat Res* 422: 82–87.

Ling, G., Bandak, F., Grant, G., Armonda, R., and Ecklund, J. 2008. Neurotrauma from explosive blast. In Elsayed, N. M., and Atkins, J. L. (Eds.), *Explosion and blast injuries: effects of explosion and blast from military operations, industrial accidents, and acts of terrorism.* Boston: Elsevier Academic Press.

Sattin, R. W., Sasser, S. M., Sullivent III, E. E., and Coronado, V. G. 2008. The epidemiology and triage of blast injuries. In Elsayed, N. M., and Atkins, J. L. (Eds.), *Explosion and blast injuries: effects of explosion and blast from military operations, industrial accidents, and acts of terrorism.* Boston: Elsevier Academic Press.

Schwartz, R. B., McManus, J. G., and Sweinton, R. (Eds.). 2008. *Tactical emergency medicine,* Philadelphia: Wolters Kluwer/Lippincott Williams & Wilkins.

Tsokos, M. 2008. Pathology of human blast lung injury. In Elsayed, N. M., and Atkins, J. L. (Eds.), *Explosion and blast injuries: effects of explosion and blast from military operations, industrial accidents, and acts of terrorism.* Boston: Elsevier Academic Press.

Chapter 8

Sharp Force Injury

FORENSIC PRINCIPLES

Sharp-edged or pointed objects that cut or pierce the skin produce sharp force injuries. Even though man-made bladed instruments such as knives, axes, or machetes are the most common sharp-edged weapons, ordinary household utensils such as ice picks, forks, industrial tools such as screwdrivers and box cutters, or edged or pointed objects of materials such as glass or wood may be used to create these wounds. Deaths from sharp force injury are the second largest group of homicidal deaths in the United States following gunshot wounds (Davis 1998). Wounds from sharp objects are customarily divided into four categories: stab wounds, incised wounds or cuts, chop or slash wounds, and therapeutic or diagnostic wounds (Di Maio and Di Maio 2001). The mechanism of death from sharp force injury is almost always hemorrhaging from massive blood loss, cardiac tamponade, or hemopneumothorax. Less commonly, air embolism or asphyxia from blood in airways may cause death. Delayed death as a consequence of sharp force injury may result from infection or sepsis (Webb et al. 1999).

Wound Classification

Stab and puncture wounds are deeper than their width at the skin surface and are produced by sharp or pointed instruments that have been forced into the body. The amount of force necessary to inflict a stab wound is related to the shape and sharpness of the weapon. Skin is the most resistant soft tissue in the body to stab wounds. Once the resistance of the skin is overcome, very little force is necessary for further penetration of the weapon (Davis 1998). The majority of stab wound fatalities are homicides. Suicides and accidents account for a small number of cases (Karger et al. 2000). In homicides, stab wounds are generally widely scattered over the body, with the life-threatening wounds in the chest or abdomen. Death is usually rapid and due to exsanguination (Di Maio and Di Maio 2001). Multiple grouped wounds may be indicative of rage. Defensive stab wounds are typically located on the hands and arms but may also be found on the legs if the victim is supine and flexes the legs into a defensive posture. Suicidal stabbing

often involves the mid or left chest. Commonly, there are multiple wounds, most of which are minimally penetrating (often called *hesitation* wounds), and only one or two deeply penetrating wounds. Accidental stab wounds are uncommon and occur from an impaling injury caused by a fall or traffic accident. Any pointed object such as a pipe or rod may be the penetrating object.

Incised wounds or cuts are wider on the skin surface than they are deep (Figure 8.1). They occur when a sharp edge is drawn over the skin. Incised wounds are lethal when they are multiple or involve a major vessel. Incised wounds and stab wounds are often found together. Homicidal incisional wounds are characterized by short slashes and swipes of varying depths. When a victim is attacked from the front, the wounds are seen on the anterior surface of the body. If the victim is attacked from behind, incisional wounds will often be seen on the anterior neck. Typically, they begin under the ear and extend downward and then upward, ending lower on the opposite side from where the wound began (Davis 1998). In suicide or attempted suicide, the individual will typically have incised wounds on the arms, neck, or both. Right-handed individuals usually cut the left wrist or forearm and vice versa, but this is highly variable. Self-inflicted incised wounds often have hesitation marks, which are superficial incised wounds adjacent to, overlying, or in continuation with the fatal wound (Di Maio and Di Maio 2001).

Heavy weapons that have a cutting edge, such as axes, hatchets, meat cleavers, and machetes, produce chop wounds. Chop wounds have a pattern of sharp force injury with blunt force components. Therefore, a combination of an incised wound with surrounding marginal abrasion and associated fractures or deep grooves in the underlying bone is characteristic of a chop injury (Davis 1998).

Medical personnel produce therapeutic and diagnostic wounds during treatment of a patient. Common examples are thoracotomy and laparotomy incisions, surgical stab wounds for the insertion of laparoscopic or thoracoscopic instruments, drainage tubes and catheters, tracheostomies, and cut downs for vascular access. Occasionally,

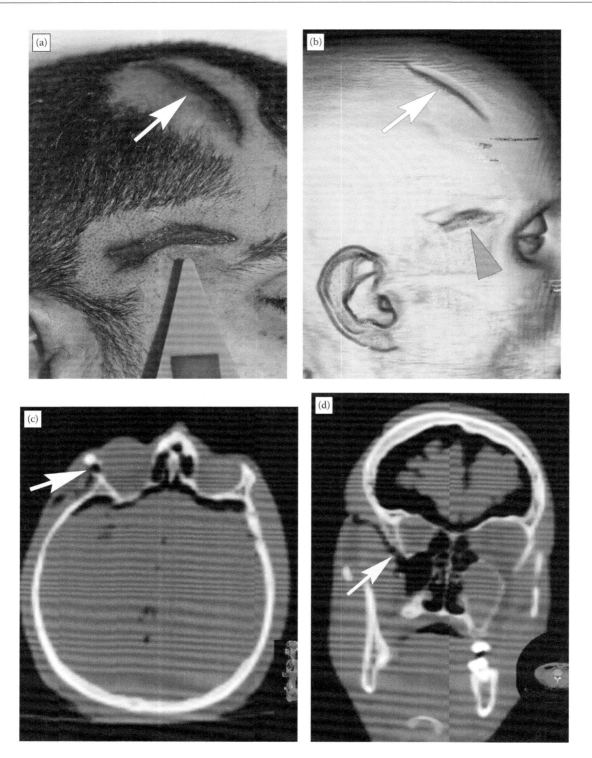

FIGURE 8.1   Stab and incised wounds of the head. (a) Autopsy photograph shows a stab wound to the right frontal temporal region (orange marker) and an incised wound of the right frontoparietal scalp (white arrow). The length of the stab wound is greater than the width of the knife blade because of blade movement at the time of injury. In contrast, the length of the incised wound exceeds its depth. The margins of the wound are sharp, and the wound lacks bridging tissue, which distinguishes it from a laceration. (b) Three-dimensional MDCT shows the wounds using a surface-rendering algorithm. The image is limited in defining wound characteristics and lacks the detail provided by photography. (c, d) Axial and coronal MDCT show that the stab wound track is directed anterior and downward. The axial image shows that the knife entered and fractured the lateral wall of the right orbit (arrow in c). The coronal image shows the downward direction of the wound track, which enters the right maxillary sinus (arrow in d).

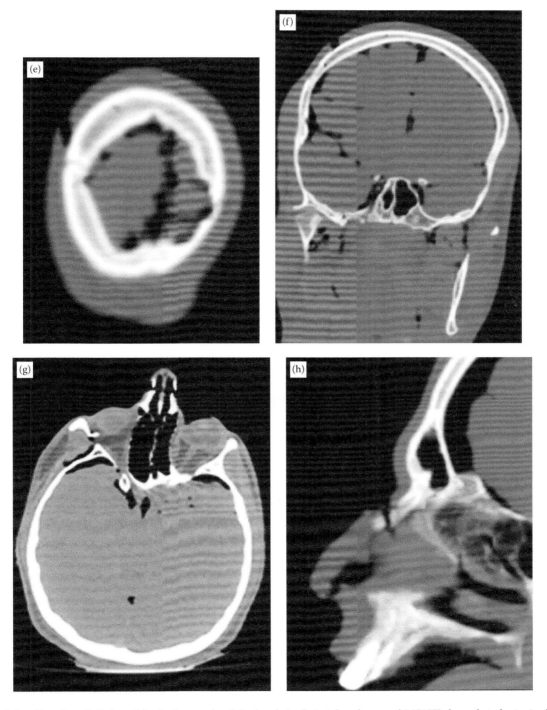

FIGURE 8.1 (*Continued*) Stab and incised wounds of the head. (e, f) Axial and coronal MDCT show that the incised wound is through the scalp but is not penetrating bone. (g, h, i) Axial, sagittal, and three-dimensional surface MDCT of an incised wound on the nose show the wound has penetrated the nasal bone.

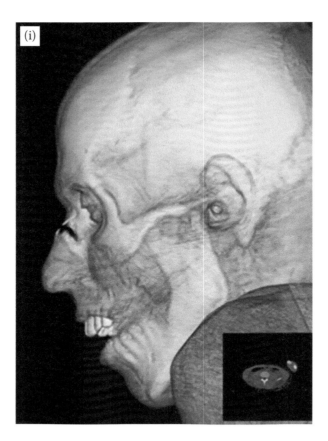

FIGURE 8.1 (*Continued*) Stab and incised wounds of the head. (g, h, i) Axial, sagittal, and three-dimensional surface MDCT of an incised wound on the nose show the wound has penetrated the nasal bone.

inflicted wounds can be confused with therapeutic wounds. This may occur when an inflicted wound is converted to therapeutic use. For example, a chest tube may be placed through a stab wound of the chest during resuscitation, or a stab wound of the abdomen may be incorporated into a laparotomy incision. To avoid misinterpretation, therapeutic tubing should not be removed prior to sending a body to the medical examiner, and medical and surgical records should be reviewed prior to autopsy (Di Maio and Di Maio 2001).

Autopsy Findings

Wounds from sharp-edged objects may initially appear to be similar to lacerations. Sharp force wounds have well-defined edges and often lack bruising and abrasion and typically do not have foreign material in or around the wound (Figure 8.1). In contrast, lacerations have irregular or ragged edges and may have associated abrasion, bruising, and foreign material. An important distinguishing characteristic of sharp force wounds is the absence of bridging tissue in the depth of the wound because a sharp force object cut tissue. Bridging tissue is the result of tissue tearing. Therefore, if connective tissue bridges are present in the depth of the wound, it is more likely a laceration associated with a blunt force injury.

At autopsy, the medical examiner measures and diagrams the location of each wound. Wounds are measured before and after reapproximating the edges of the wound. The size, shape, anatomic location, and orientation of the wound are described as well as the appearance of the edges (blunt or sharp). The presence of associated marks is noted because they may reflect the physical characteristics of the weapon. For example, the guard or hilt of a knife or tool may leave an impression adjacent to the wound. Following external examination, visceral and vascular injuries are identified during dissection, and a determination of lethal injury is made (Figure 8.2).

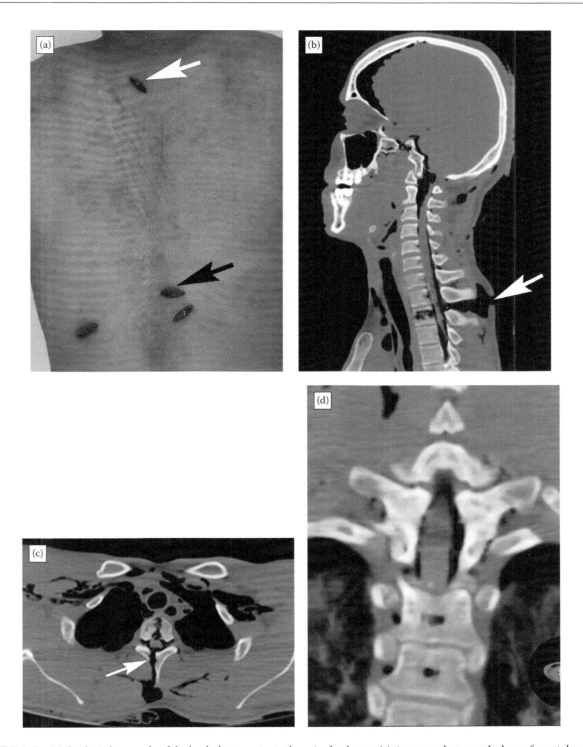

FIGURE 8.2 Multiple stab wounds of the back that penetrate the spinal column. (a) Autopsy photograph shows four stab wounds. Two are located in the midline (white and black arrows). (b, c, d, e, f, g) Sagittal, axial, and coronal MDCT show the superior midline wound track (arrow). The wound track has a vertical orientation and penetrates the second thoracic vertebral body passing through the right lamina, spinal canal, and body. The spinal cord is intact. Three-dimensional reconstructions illustrate the bone defects (arrows).

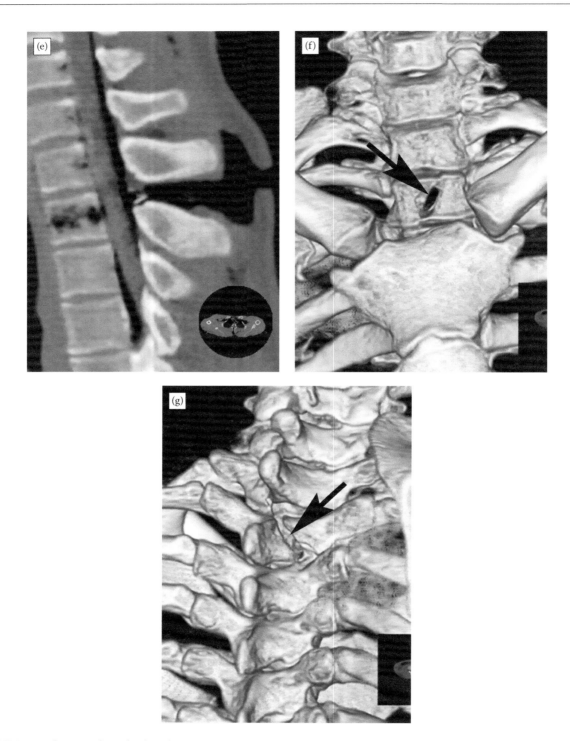

FIGURE 8.2 (*Continued*) Multiple stab wounds of the back that penetrate the spinal column. (b, c, d, e, f, g) Sagittal, axial, and coronal MDCT show the superior midline wound track (arrow). The wound track has a vertical orientation and penetrates the second thoracic vertebral body passing through the right lamina, spinal canal, and body. The spinal cord is intact. Three-dimensional reconstructions illustrate the bone defects (arrows).

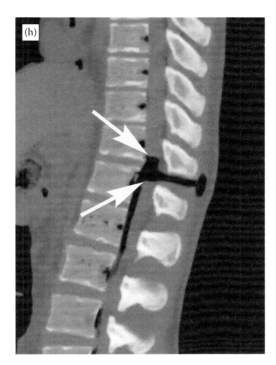
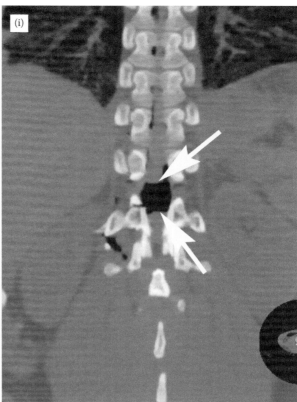

FIGURE 8.2 (*Continued*) Multiple stab wounds of the back that penetrate the spinal column. (h, i) Sagittal and coronal MDCT of the inferior midline stab wound show that the wound has a horizontal orientation and passes between the posterior vertebral elements and into the spinal canal, severing the spinal cord. The ends of the cord are retracted (arrows). There is no break in the skin surface, which is likely due to closure of the wound from the supine positioning of the body. On the coronal image, there is a small, linear collection of air located to the inferior right of the midline wound. This is an adjacent stab wound.

The edges of stab wounds are typically sharp (Figure 8.1 through Figure 8.3). The configuration of stab wounds is variable. The force and movement of the weapon, angle of thrust, elasticity of the skin, and degree of tissue compression affect wound configuration. The natural pattern of collagen and dermal elastic fibers in the dermal layer of the skin (Langer's lines) affects the separation of the edges of a stab wound. If the wound is parallel to Langer's lines, it will be slit-like; if it is perpendicular, it will gape. The movement of the weapon within the wound at the time of injury will affect the appearance of the wound. For example, if a knife is twisted within the wound, the wound will be changed from slit-like to a nonlinear irregular defect (Davis 1998). The length of a stab wound will be equal to, less than, or greater than the width of the knife, and the depth of the wound may also be equal to, less than, or greater than the length of the knife (Bauer and Patzelt 2002). Consequently, medical examiners do not try to

give an estimate of the width or length of the offending blade based upon their determination of the depth of the wound in the body (Davis 1998, Di Maio and Di Maio 2001).

The classic incised wound is clean-cut and free of abrasion or contusion. A dull, irregular, or nicked cutting edge may produce a wound that has irregular, contused, or abraded margins. However, there will be no bridging tissue in the depth of the wound. Incised wounds, like stab wounds, will tend to separate with a degree of separation that is related to the orientation of the wound along Langer's lines. Incised wounds often begin superficially, increase in depth, and end superficially. A wrinkle wound is a variation of an incised wound that is characterized by a string of cuts with gaps of normal tissue between the cuts. This is produced when a cutting edge is drawn over skin with folds and the cuts are made in a

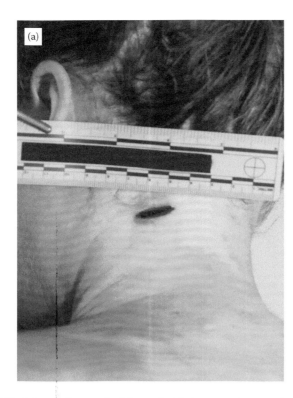
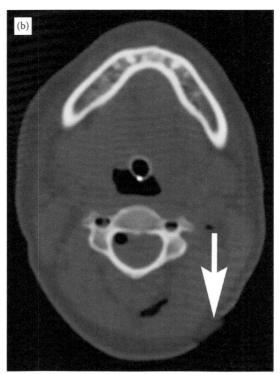

FIGURE 8.3   Stab wound of the neck. (a) Autopsy photograph shows a horizontal stab wound on the left posterior neck below the hairline. (b) Axial MDCT shows entry wound (arrow) and soft tissue air in the track. An endotracheal tube is in place.

skip-like manner on the crests of the skin folds (Di Maio and Di Maio 2001).

Although most chop wounds appear incised, the combination of cutting and crushing produced by a heavy weapon results in an injury that has characteristics of both incised wounds and lacerations. Fractures may accompany chop wounds. The type and severity of an associated fracture vary with the type of weapon, anatomic location, and manner in which the blow was delivered to the body. For example, tangential chopping wounds of the skull may excise disc-shaped fragments of bone. Chopping weapons produce striations in the bone, which are unique to a specific weapon (Humphrey and Hutchinson 2001, Thali et al. 2003). Cleavers, machetes, and axes will each produce different patterns when they impact on bone.

Therapeutic wounds are usually easily differentiated from inflicted wounds by their appearance, anatomic location, and medical history. They are most often the result of emergency treatment. In the chest, emergency thoracotomy incisions are usually long, deep incisions in an intercostal space. In the abdomen, laparotomy incisions are most common in midline of the anterior abdominal wall. Stab-like therapeutic wounds are used for placement of chest tubes and drainage catheters and may be found in the chest or abdomen.

RADIOLOGIC PRINCIPLES
Goals of Imaging

Traditionally, conventional radiography was considered an important component in the forensic assessment of sharp force injury. Radiography was used to help identify and aid recovery of broken knife blades and to help differentiate small round stab wounds (e.g., ice pick wounds) from small-caliber ballistic wounds (Davis 1998). More recently, the availability of cross-sectional imaging has expanded the role of imaging in the assessment of these wounds. Stab wound tracks and tissue and organ injury can be assessed with multidetector computed tomography

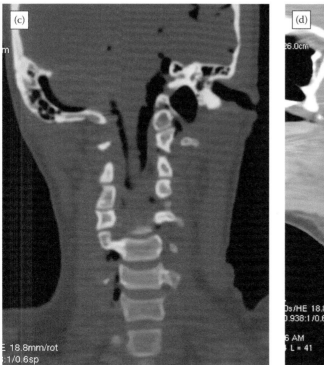

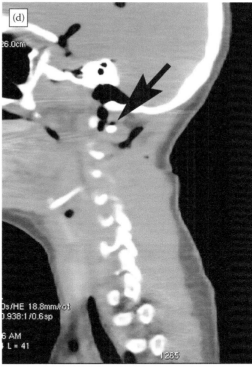

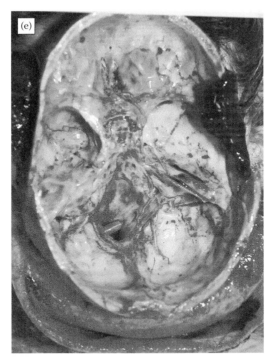

FIGURE 8.3 (*Continued*) Stab wound of the neck. (c) Oblique coronal MDCT shows track coursing medial and upward. (d) Oblique sagittal MDCT shows air collections in the track to include a collection above the posterior arch of C1 (arrow). (e) Autopsy photograph with a rod in the wound track shows the spinal canal was entered. This case also shows intracranial and paraspinal air secondary to decomposition, which can be difficult to distinguish from pathologic air collections.

(MDCT) prior to dissection. This is particularly useful in wounds that are located in anatomic areas that are more difficult to dissect, such as those located in and around the spine (Figures 8.2 and 8.3).

Imaging Findings

The majority of weapons used to produce sharp force injury are metallic and unlikely to break apart in soft tissue. However, the tip of a knife blade may break if it is embedded in bone. When this occurs, identification and recovery of the fragment is forensically important. The recovered fragment can aid in more precise identification of the weapon, and if a weapon is recovered, a physical match can be made (Di Maio and Di Maio 2001). Knife blade fragments are metallic attenuation on radiographs and MDCT. On radiography, the location of the fragment should correlate with the expected location of the wound path based on the skin entry site. On MDCT, the wound track may be visualized and lead to the metallic fragment. On rare occasions, a knife is found embedded in the body. In these instances, orthogonal radiographs can be used to estimate the depth of the knife in the body. Although MDCT images may be degraded by streak artifact from a large metallic object, they may still provide useful three-dimensional reconstructions that may be of benefit later in legal presentations (Bauer and Patzelt 2002, Karlins et al. 1992, Thali et al. 2002).

Skin surface mapping and characterization of wounds is a major focus of the forensic autopsy in sharp force injury. When there are multiple stab wounds, it may be helpful to place a metallic marker on the skin entry site after the pathologist has completed the external examination, and then repeat MDCT imaging so direct correlation is facilitated. The shape and edge patterns of the wound that are studied visually by the forensic pathologist have a variable appearance on MDCT. Because currently available software does not provide the skin detail necessary for accurate characterization of the wound surface, it is not recommended that MDCT be used to make determinations of blade type, such as estimating whether or not a knife was single edged, double edged, or serrated (Figure 8.1). The visibility of a wound on MDCT is dependent on the orientation and position of the body. A gap in the skin surface from an open wound can close when the wound is dependent.

Air is the key finding to identify a wound track from a sharp-edged object. This is the same principle used to identify gunshot and projectile wound tracks. On the skin surface, a break in the continuity of the skin is outlined by air, and the track in the body is visible if air is carried into soft tissue or released from a gas-containing organ such as lung or bowel (Figure 8.2 through Figure 8.5). Air from venous air embolism may be seen in the right heart and venous structures following stab wounds to the neck or wounding of any major vein that permits air to enter the venous system (Dirnhofer et al. 2006).

Although hemorrhage is the most common cause of death in sharp force injury, bleeding may be primarily external and consequently not visible on MDCT. Internal hemorrhage is visible in the spaces where blood accumulates in sufficient volume. Hemopneumothorax, hemoperitoneum, perinephric hematoma, and subcapsular hemorrhage in intra-abdominal organs are usually readily identified on MDCT (Figures 8.6 and 8.7). Injuries to the heart are likely to cause hemopericardium with cardiac tamponade (Dirnhofer et al. 2006).

Stab wounds are recognized on axial images by breaks in skin surface contour and the presence of subcutaneous air (Figures 8.2, 8.3, and 8.5). By viewing sequential images on a workstation or using multiplanar reconstructions, it may possible to determine the orientation and approximate length of the wound at the skin surface. It may also be possible to estimate the depth of the wound if the wound track contains gas or if there is adjacent bone injury (Figures 8.3, 8.6, and 8.7). These measurements must be done in the plane of the wound track. However, it is not recommended to correlate these measurements with the physical measurements or the measurements of a suspected weapon. All of the skin variables that create the lack of correlation between weapon dimensions and physical wound measurements on the body apply when images are analyzed. Consequently, the radiologist should not state that a particular weapon is the type used in the case.

Wound tracks in stab injury become visible on MDCT when air enters the wound track, bone is damaged, or the track fills with blood. There is generally less air in stab wounds than in gunshot wounds, particularly as depth increases, because the wound does not experience cavitation. The tissue surrounding the wound may quickly close

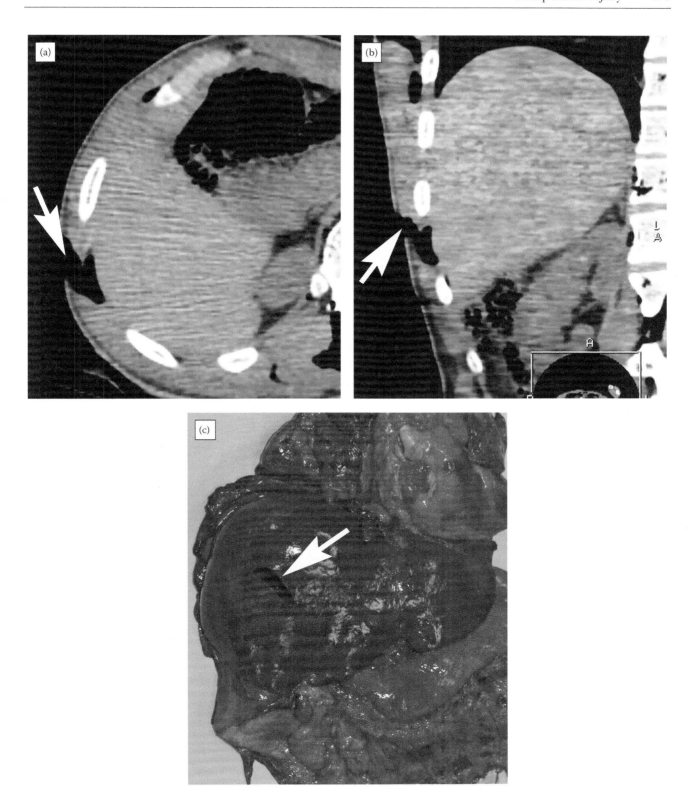

FIGURE 8.4   Stab wound of the abdomen. (a, b) Axial and coronal MDCT show air in the stab wound track in the abdominal wall (arrows). There is no evidence on the image that the stab wound penetrates the liver. (c) Autopsy photograph shows the wound track penetrated the right lobe of the liver (arrow). The wound is not evident on MDCT because the wound edges are closed and no air entered the liver parenchyma.

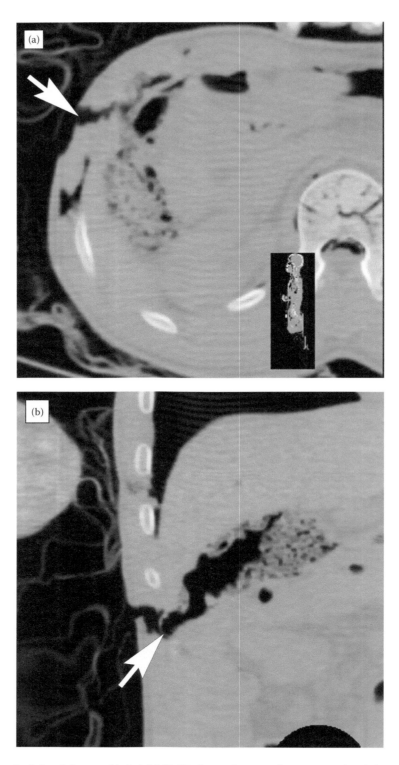

FIGURE 8.5   Stab wound of the abdomen. (a) Axial MDCT shows the wound penetrates the abdominal wall (arrow). There is a second air collection in the abdominal wall posterior to the stab wound which represents a second stab wound, which is less apparent because margins of the wound are closed at the skin surface. (b) Coronal MDCT clarifies the depth of the stab wound and damage to tissue by revealing that the wound penetrated the hepatic flexure of the colon (arrow).

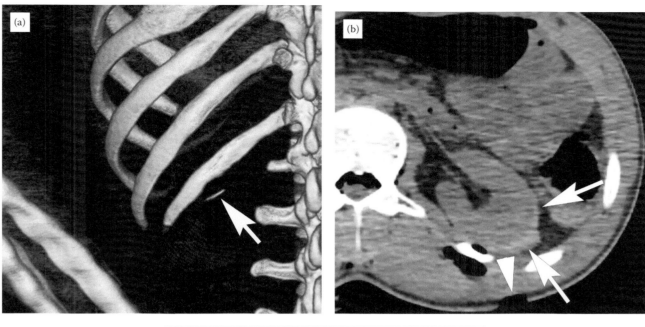

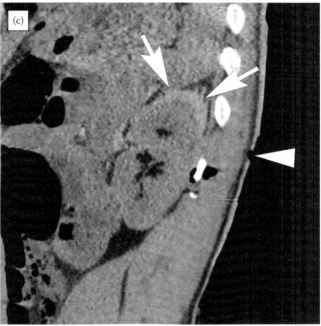

FIGURE 8.6   Stab wound of the left flank showing that the wound track can be determined from the combination of bone and soft tissue findings. (a) Three-dimensional MDCT of the left rib cage shows a linear fragment of bone cut from the lower margin of the posterior 12th rib (arrow). (b, c) Axial and sagittal MDCT show a rim of high-attenuation fluid around the left kidney which represents perinephric hemorrhage (arrows). This is evidence that the stab wound entered the left kidney. The intrarenal wound track is not apparent because the edges are closed and no air has entered the kidney. There is a defect on the skin surface (arrowheads), air collection around the left 12th rib, and the bone fragment below the rib; these findings indicate the knife had a downward direction.

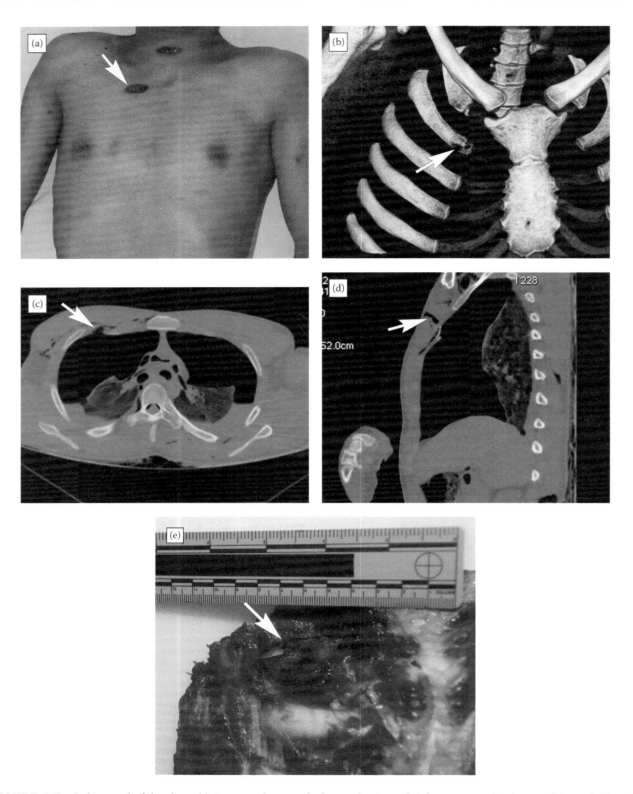

FIGURE 8.7 Stab wound of the chest. (a) Autopsy photograph shows a horizontal right parasternal stab wound (arrow). There is another wound in the midline of the neck. (b) Three-dimensional MDCT of the thorax reveals a defect in the second right rib adjacent to the costochondral junction (arrow). (c, d) Axial and sagittal images of the thorax show the wound track in the right anterior chest wall (arrows). There is bilateral pneumothorax and hemothorax. (e) Autopsy photograph of the chest plate shows the stab wound through the costal cartilage and second interspace on the right (arrow).

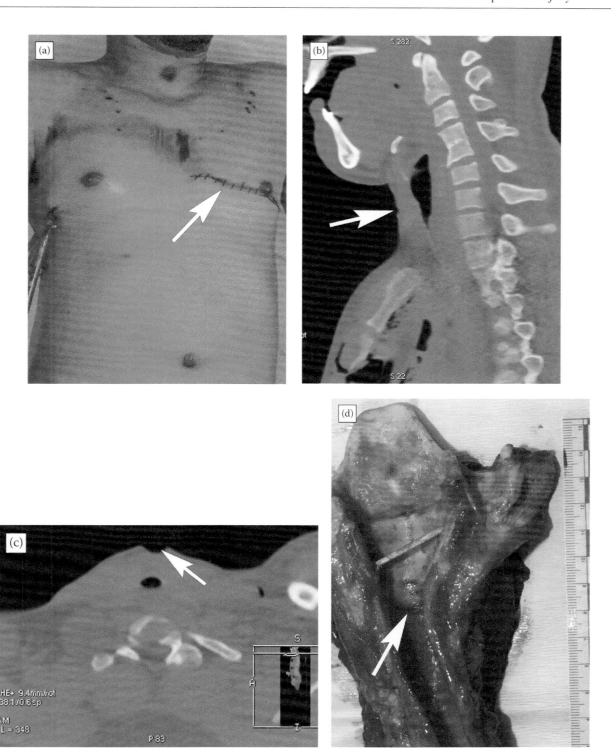

FIGURE 8.8 Surgical wounds of the chest and neck from a thoracotomy with superficial closure and an attempted tracheotomy. (a) Autopsy photograph shows a sutured left thoracotomy (arrow), right chest tube, sternal abrasion, and transverse neck wound. (b, c) Sagittal and axial images of the neck reveal a superficial anterior wound (arrows) but no wound track in the underlying soft tissues. Incidentally, there is a cervical vertebral fusion anomaly of C4-C5. (d) Autopsy photograph of the trachea from posterior reveals a horizontal wound entering the trachea (arrow) not seen on MDCT.

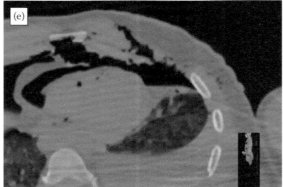

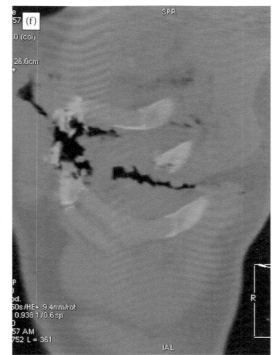

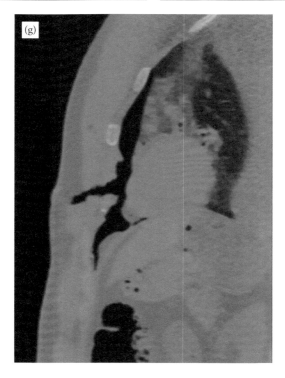

FIGURE 8.8 (*Continued*) Surgical wounds of the chest and neck from a thoracotomy with superficial closure and an attempted tracheotomy. (e) Axial image of the left hemithorax shows wound tracks extending through the anterior chest wall into the pleural space. The pericardium is open; there is a hemopneumothorax; and there is a wavy skin surface contour, which is from the skin sutures of the thoracotomy incision. (f, g) Oblique maximal intensity projection and sagittal MDCT show that the surgical incision follows an intercostal space.

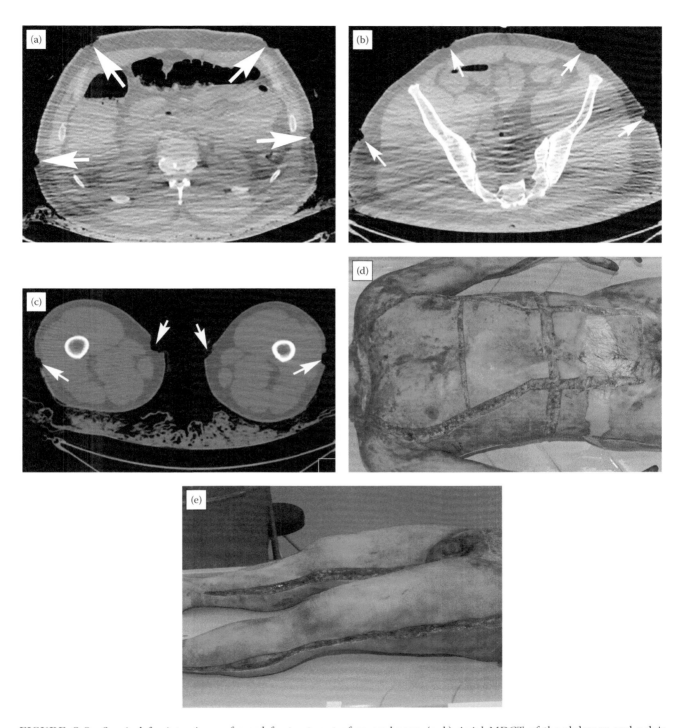

FIGURE 8.9 Surgical fasciotomies performed for treatment of severe burns. (a, b) Axial MDCT of the abdomen and pelvis shows bilateral anterior and lateral fasciotomy defects in the skin (arrows). (c) Axial MDCT of the thigh at the mid-femur level shows bilateral medial and lateral fasciotomy defects (arrows). (d, e) Autopsy photographs show the linear fasciotomy incisions on the torso and legs.

when the blade is withdrawn, making the wound invisible on MDCT (Figure 8.4). The interaction of a sharp edge weapon with bone can produce a cut defect, chip fragment (Figure 8.7), hole, or fracture (Figure 8.2). The type of bone injury depends on the knife blade shape and type of the bone stabbed. The presence of bone–blade interaction is a good indicator of the wound track direction and aids in the approximation of wound depth (Bauer and Patzelt 2002) (Figure 8.6). MDCT is valued for its ability to show wound changes in areas that are difficult for the medical examiner to access at autopsy (Figure 8.1 through Figure 8.3).

The shift of organs from position at time of attack to position on the postmortem MDCT is a critical factor in determining wound track. For example, normally expanded lungs place the heart in a different anatomic position than it is on a supine postmortem image with bilateral pneumothorax (Figure 8.7). The wound track and organ position in stab wounds of the abdomen will be affected by whether the victim was sitting or standing at the time of injury (Figure 8.4).

Incised wounds can be recognized because their length exceeds their depth. This can be readily appreciated when viewing MDCT images on a workstation. Using the multiplanar capability of MDCT, the imaging plane can be aligned perpendicular to the long axis of the wound to show that the wound is free of bridging tissue in the depth of the wound. The lack of bridging tissue distinguishes a knife wound from a laceration. Incised wounds, like stab wounds, will exhibit varying degrees of separation related to the orientation of the wound and the position of the body at the time of the scan. For example, wounds on the back will often be closed when the body is scanned supine. We have found that incised wounds on the face and scalp are easier to visualize than those on the body.

Because chop wounds are the combination of cutting and crushing by a heavy weapon, the MDCT findings of chop wounds have soft tissue characteristics of both incised wounds and lacerations. Chop wounds frequently have associated fractures. Chop wound fractures vary with the type of weapon, location on the body, and way in which the blow was delivered. For example, tangential chopping wounds of the skull may cut off discs of bone. Chopping weapons that pass through bone can impart striations on the bone that are unique to the type of weapon: cleaver,

machete, or axe (Humphrey and Hutchinson 2001, Thali et al. 2003).

Surgical wounds are evident when tubes or drains remain in place, metallic staples are used, devices are present, or sponges remain in the body. A clue to surgical wounds is their location and direction (Figures 8.8 and 8.9). For example, thoracotomies typically follow an intercostal space, and chest tube wounds are usually placed in the anterior axillary line. When death occurs after a surgical procedure and the wounds have been closed, it is easy to fail to identify the incisions on MDCT if there has been standard layer-by-layer closure. There are occasions when tubes or drains have been removed before (or after) death and also times when a stab wound has been incorporated into a surgical incision. When death occurs during resuscitative surgery, surface suturing may be used to close the incisions, and the deeper layers are left unopposed. We observed this in thoracotomies (Figure 8.8) and laparotomies. The radiologist will experience the same difficulties as the pathologist in distinguishing therapeutic wounds from inflicted wounds when the medical records are absent or incomplete.

CONCLUSIONS

Wounds caused by sharp-edged objects are to be characterized by MDCT using body surface renderings in combination with multiplanar reconstructions displayed to show the wound track. Damage to bone and soft tissue structures allow estimation of wound depth and direction. MDCT is also useful to detect wound penetration into areas of the body that are difficult to reach during routine autopsy. It should be emphasized that no conclusion should be offered regarding a specific type of weapon based on MDCT wound characteristics.

REFERENCES

Bauer, M., and Patzelt, D. 2002. Intracranial stab injuries: case report and case study. *Forensic Sci Int* 129: 122–127.

Davis, G. J. 1998. Patterns of injury. Blunt and sharp. *Clin Lab Med* 18: 339–350.

Di Maio, V. J. M., and Di Maio, D. J. 2001. *Forensic pathology*, Boca Raton, FL: CRC Press.

Dirnhofer, R., Jackowski, C., Vock, P., Potter, K., and Thali, M. J. 2006. VIRTOPSY: minimally invasive, imaging-guided virtual autopsy. *Radiographics* 26: 1305–1333.

Humphrey, J. H., and Hutchinson, D. L. 2001. Macroscopic characteristics of hacking trauma. *J Forensic Sci* 46: 228–233.

Karger, B., Niemeyer, J., and Brinkmann, B. 2000. Suicides by sharp force: typical and atypical features. *Int J Legal Med* 113: 259–262.

Karlins, N. L., Marmolya, G., and Snow, N. 1992. Computed tomography for the evaluation of knife impalement injuries: case report. *J Trauma* 32: 667–668.

Thali, M. J., Schwab, C. M., Tairi, K., Dirnhofer, R., and Vock, P. 2002. Forensic radiology with cross-section modalities: spiral CT evaluation of a knife wound to the aorta. *J Forensic Sci* 47: 1041–1045.

Thali, M. J., Taubenreuther, U., Karolczak, M. et al. 2003. Forensic microradiology: micro-computed tomography (Micro-CT) and analysis of patterned injuries inside of bone. *J Forensic Sci* 48: 1336–1342.

Webb, E., Wyatt, J. P., Henry, J., and Busuttil, A. 1999. A comparison of fatal with non-fatal knife injuries in Edinburgh. *Forensic Sci Int* 99: 179–187.

Chapter 9

Death by Drowning and Bodies Found in Water

FORENSIC PRINCIPLES

The forensic investigation of a body recovered from water begins with the consideration that death may have occurred as a result of drowning, natural disease, or injury before falling or being placed in the water, or from natural disease or injury while in the water. The differentiation between these causes of death may be difficult because almost all bodies recovered from water will have cutaneous signs of immersion and may have fluid within the tracheobronchial tree or stomach. Consequently, drowning is one of the most difficult causes of death for a forensic pathologist to render. The longer a body is in the water, the more indeterminate the diagnosis will become. The presence of decomposition or marine animal predation may make determination of the cause of death more complex or impossible in some cases. Of greatest importance, the diagnosis of drowning is a diagnosis of exclusion, because there is no specific or conclusive test to diagnose drowning. The most reliable method to determine that a death was due to drowning is to compare the anatomic findings at autopsy with the scene investigation and exclude other causes of death.

Physiology of Drowning

Drowning was defined by the World Congress on Drowning as a process that results in primary respiratory impairment from submersion or immersion in a liquid medium (Idris et al. 2003). When a drowning victim's airway lies beneath the surface of the liquid, the victim voluntarily holds his or her breath. Liquid in the pharynx or larynx causes involuntary laryngospasm, which results in hypoxemia and a hypercarbic metabolic acidosis. As the victim's hypoxemia worsens, laryngospasm abates, which causes the victim to actively breathe in liquid. However, the amount of inhaled liquid is variable from person to person, and the changes that occur in the lungs, body fluids, blood gases, and electrolytes are also variable, depending on the duration, volume, and composition of inhaled liquid (Idris et al. 2003).

Consciousness is usually lost within 3 minutes of submersion (Pearn 1985). Death occurs from tissue hypoxia. The terminal event is circulatory collapse from cardiac arrhythmia with or without a terminal seizure.

The physiologic basis of hypoxemia and the formation of pulmonary edema in drowning are dependent on the type of water inhaled. In saltwater drowning, fluid is drawn into the alveoli from plasma, because there is hypertonic saltwater in the alveoli (Modell et al. 1967). In contrast, freshwater alters the surface tension properties of surfactant, which causes some alveoli to become unstable or collapse, altering the ventilation–perfusion ratio and forming intrapulmonary shunts (Modell 1993). Large amounts of freshwater pass through the alveolar capillary bed causing hypervolemia and electrolyte dilution with rapid redistribution of fluid and formation of pulmonary edema (Modell et al. 1966). Other contributing factors to the formation of pulmonary edema in all types of drowning include pulmonary hypertension and cerebral hypoxia (Modell 1978). If particulate matter is present in the water, it may mechanically obstruct small bronchi and bronchioles.

The term *dry drowning* has been used in the past to distinguish between victims who aspirate fluid into the lungs during drowning and those who do not. Drowning victims with significant fluid aspiration (wet drowning) have heavy, edematous lungs at autopsy. In contrast, dry drowning victims have no significant signs of fluid penetration into the lower airways and normal or low-weight lungs with hyperinflation at autopsy. It has been hypothesized that dry drowning victims have little or no penetration of liquid below the larynx because of profound laryngospasm. The existence of dry drowning has been disputed by some and remains a controversial topic. Some authors suggest that making the diagnosis of drowning without evidence of aspirated water is risky, because some causes of sudden death such as cardiac arrhythmia cannot be confirmed postmortem (Modell et al. 1999). In addition, recent literature suggests that lung

183

weights are not reliable criteria for the distinction between types of drowning and that the incidence of low lung weight in drowning is much less (approximately 2% of drowning cases) than previously suggested (Lunetta et al. 2004).

Autopsy Findings

The autopsy diagnosis of drowning is based on scene investigation, circumstances of death, and anatomic findings at autopsy that are supportive of drowning. Drowning is a diagnosis of exclusion and can only be made after complete autopsy with toxicology, because the anatomic findings of drowning may overlap those of other causes of death. For example, death in water may be due to any number of natural causes, such as sudden cardiac death, preexisting cardiac disease, pulmonary embolism, stroke, or subarachnoid hemorrhage. Hypothermia or exhaustion may also play a role in drowning deaths. On the other hand, victims of natural death, impaired judgment from alcohol or drug use, drug overdose, or homicide may fall or be thrown into water. In all of these scenarios, fluid may be present in the airways and lungs from pulmonary edema of various etiologies or from agonal respirations underwater.

Bodies recovered from water often display signs of immersion on external examination. Signs of immersion indicate that the body has been in water, but they are not specific for the cause of death. The degree and extent of cutaneous changes of immersion depend on the temperature of the water and the length of time a body was in the water. There is often maceration of the skin where the skin surface becomes wrinkled, pale, and sodden. This is usually most pronounced on the portions of the body not protected by clothes, such as the hands and feet (Figure 9.1). Cutis anserine, also called *goose flesh*, is the appearance of dimpling of the skin from contraction of erector pilae muscles of each hair follicle and may be seen in bodies retrieved from cold water. Mud, silt, or sand may be present on the skin, clothing, and oral and nasal cavities from victims recovered from water heavily laden with particulate matter. One of the most important components of the external examination is evaluating the body for evidence of antemortem blunt or penetrating trauma. Intentional trauma, suicide, homicide, child abuse, and impairment from alcohol or drug use must always be considered at autopsy. Antemortem traumatic injuries may be difficult to distinguish from postmortem injuries because the wounds may appear bloodless following prolonged immersion (Di Maio and Di Maio 2001).

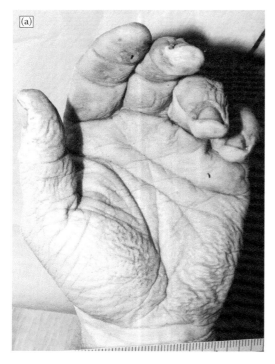 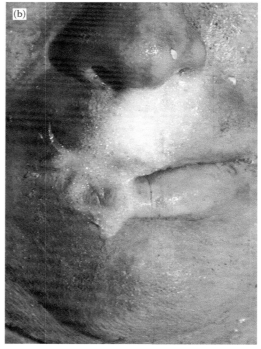

FIGURE 9.1 External autopsy findings in drowning. (a) Pale, wrinkled palm (often called *washerwoman's skin*) caused by prolonged immersion in water. (b) Froth exuding from the mouth and nostrils in a person who drowned. There is also sand on the chin and a laceration of the upper right lip that was likely from antemortem blunt trauma.

Anatomic findings at autopsy to support the diagnosis of drowning include the presence of frothy fluid in the airways or lungs, hyperinflated and congested lungs, fluid in the paranasal sinuses, watery fluid in the stomach, and dilated and engorged right-sided cardiac chambers and great vessels. The frothy fluid in the airways and lungs is a proteinaceous exudate of intra-alveolar edema and surfactant mixed with water of the drowning medium (Gordon 1972). In bodies freshly recovered from water, froth may exude from the mouth (Figure 9.1). It is typically white but may be pink or red-tinged from pulmonary hemorrhage secondary to pressure gradient changes when the body is submersed. Froth may fill portions of the tracheobronchial tree (Figures 9.2 and 9.3) or exude from

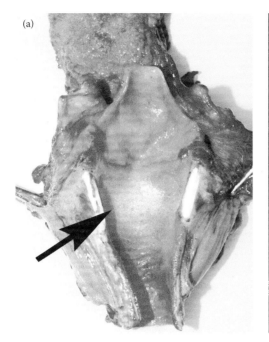

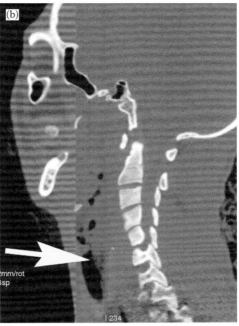

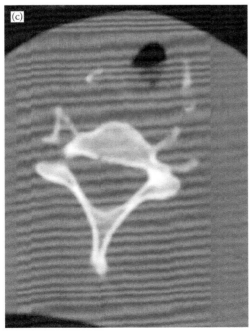

FIGURE 9.2   Airway froth. (a) White, foamy froth (arrow) in the larynx. (b, c) Oblique and axial MDCT shows laryngeal froth as the low-attenuation material admixed with rounded foci of gas (arrow) within the trachea.

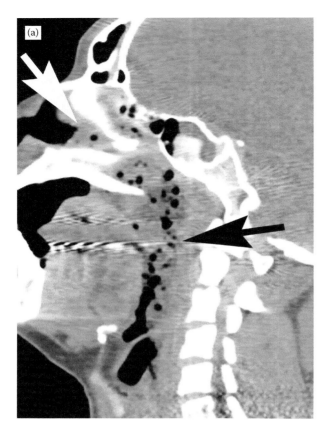

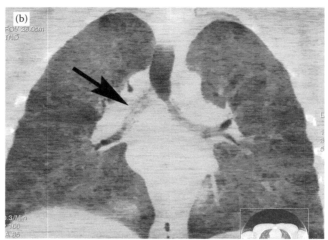

FIGURE 9.3   Airway froth. (a) Coronal MDCT shows frothy fluid filling the posterior nasal cavity (white arrow), nasopharynx, and oropharynx (black arrow). Fluid is also in the frontal and sphenoid sinuses. (b) Coronal minimum intensity projection MDCT image of the chest shows froth in the right (arrow) and left main bronchi. The trachea, bronchus intermedius, left upper lobe bronchus, and left lower lobe bronchus are well aerated, but there is no air in the more distal bronchi bilaterally.

the cut surface of the lung. Hyperinflated lungs at autopsy are defined as lungs that are expanded with air such that they fill the pleural cavity and encroach upon the mediastinum or cross the midline and touch one another when the thoracic cavity is opened and the sternum is removed. The term *kissing lungs* was applied to the appearance of hyperinflated lungs at autopsy (Figure 9.4). In some cases, the lungs are so distended that the ribs leave an impression on the surface of the lungs. Congested or edematous lungs are also characteristic and typically result in combined lung weights of greater than 1000 grams (Figure 9.5 through Figure 9.8) (Kringsholm et al. 1991). Petechial hemorrhages may be present in the lungs and on the pleura. Because inhaled and aspirated fluid enters the paranasal sinuses as it passes through the nasal cavity, the finding of sphenoid and ethmoid sinus fluid is supportive of the diagnosis of drowning. Sphenoid sinus fluid may be sampled by aspiration or unroofing of the sphenoid sinus at autopsy (Figure 9.9 through Figure 9.11). The finding

of sand or sediment in the paranasal sinuses and airways is an important marker of inhalation and aspiration of sediment-laden water (Figure 9.11 through Figure 9.13). Although many consider the findings of pleural effusion, watery fluid in the stomach, and dilated, engorged right-sided cardiac chambers and great vessels as nonspecific, they are helpful, supportive findings when other anatomic findings of drowning are present (Figures 9.14 and 9.15). Many authors have sought to find more specific findings to diagnose drowning, such as the detection of diatoms, serum biochemical analysis, and comparison of organ weights, but these have not been shown to be consistently reliable (Azparren et al. 2003, Azparren et al. 1998, Hadley and Fowler 2003, Hurlimann et al. 2000).

RADIOLOGIC PRINCIPLES
Goals of Imaging

The primary goal of imaging a body recovered from water is to exclude other causes of death. Full-body multidetector

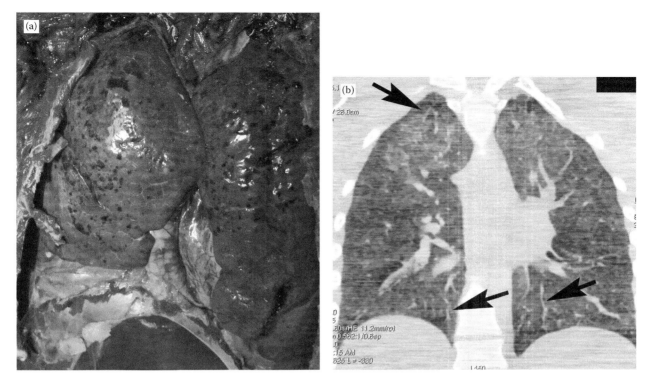

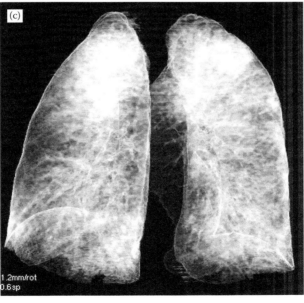

FIGURE 9.4 Hyperinflated, congested lungs in drowning. (a) Autopsy photograph of the lungs in situ shows hyperinflation. The lungs completely fill the pleural cavities and touch at midline, obscuring the anterior mediastinum. (b) Coronal MDCT of the lungs shows diffuse ground glass opacity and the thick septal lines (arrows) of pulmonary edema. (c) Three-dimensional volume-rendered image shows that the pulmonary edema produces a diffuse increase in lung opacity with more severe opacity in the apexes.

computed tomography (MDCT) is the most appropriate cross-sectional imaging technique to rapidly anatomically survey a body recovered from water for evidence of skeletal trauma, metallic fragments from projectile or gunshot injury, or gunshot wound tracks. MDCT is superior to

radiography for the detection and assessment of gunshot wound injury (Harcke et al. 2007). After images have been evaluated for evidence of antemortem trauma and disease, the images should be meticulously evaluated for anatomic findings supportive of the diagnosis of drowning.

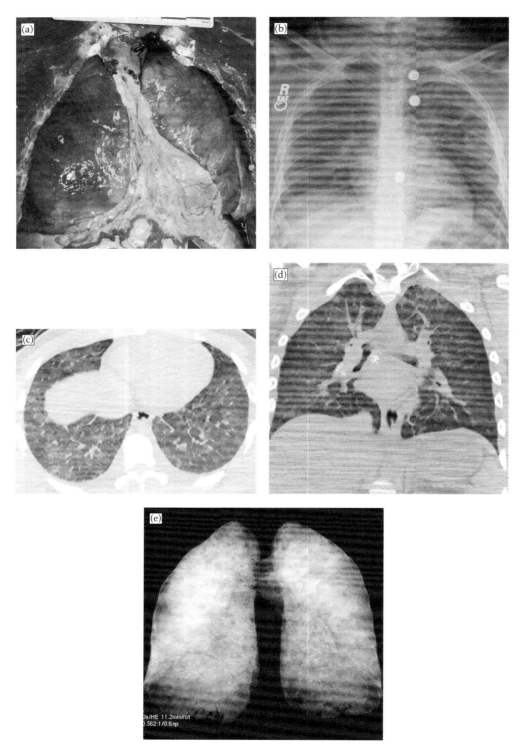

FIGURE 9.5   Congested, edematous lungs in drowning. (a) Autopsy photograph shows a mosaic pattern of pale hyperinflated areas of the lung mixed with areas that are deeper red in color. Focal areas of hypoxic vasoconstriction or bronchoconstriction cause the pale appearance. (b) Digital radiography shows fine, fluffy, alveolar pulmonary edema and the indistinctness of the hilar vasculature. The distal trachea is fluid filled. (c) Axial MDCT of the lung bases shows a mosaic pattern of severe ground glass opacity, and septal lines. (d) Coronal MDCT of the lungs shows moderate pulmonary edema evenly distributed throughout the lungs. Prominent septal lines are present in the bases and apexes. Incidentally, granulomatous calcification is in the right hilar and subcarinal lymph nodes. (e) Three-dimensional volume-rendered image of the lungs shows that the pulmonary edema produces a diffuse increase in lung opacity.

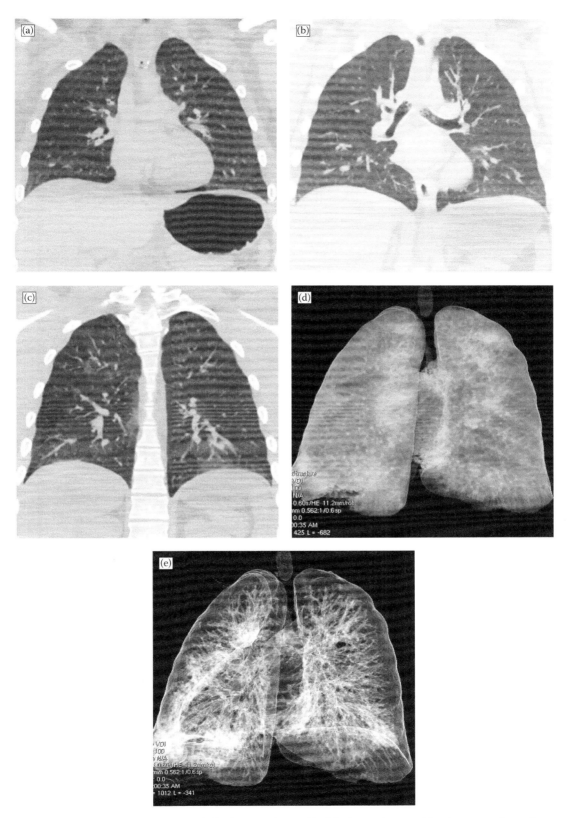

FIGURE 9.6 Pulmonary edema in drowning. (a, b, c) Coronal MDCT of the lungs shows mild pulmonary edema characterized by interlobular septal thickening and scattered, mild ground glass opacities. Froth is present in the right and left mainstem bronchi in (b). (d, e) Three-dimensional volume-rendered images show mild diffuse, symmetric pulmonary edema.

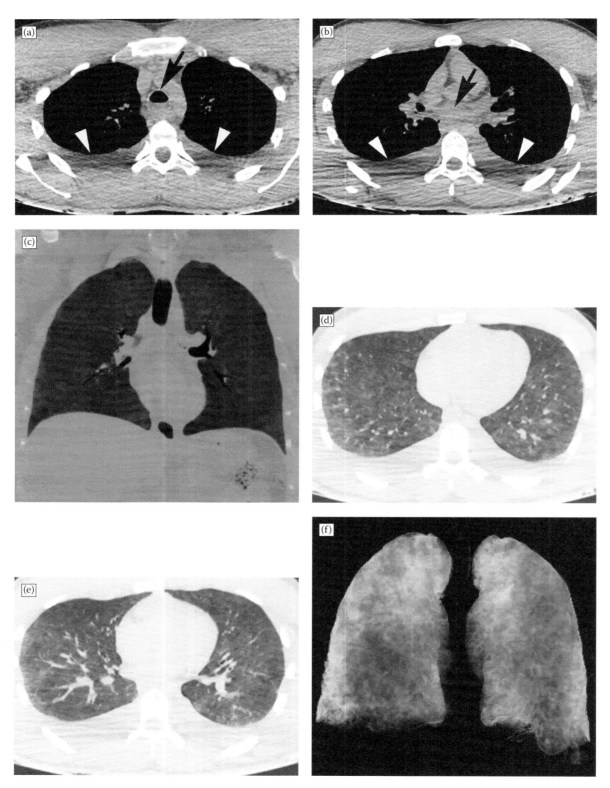

FIGURE 9.7   Freshwater drowning. (a, b) Axial MDCT images of the chest show fluid in the trachea (arrow in a) and at the carina (arrow in b). There are small, bilateral pleural effusions (white arrowheads). (c) Coronal minimum intensity projection image shows diffuse pulmonary edema and fluid opacification of the distal trachea and main stem bronchi. Note that the smaller, distal bronchi are not visualized because they are fluid filled. (d, e) Axial MDCT images show moderate diffuse ground glass opacity of pulmonary edema and bilateral effusions. (f) Three-dimensional volume-rendered image shows a diffuse edema.

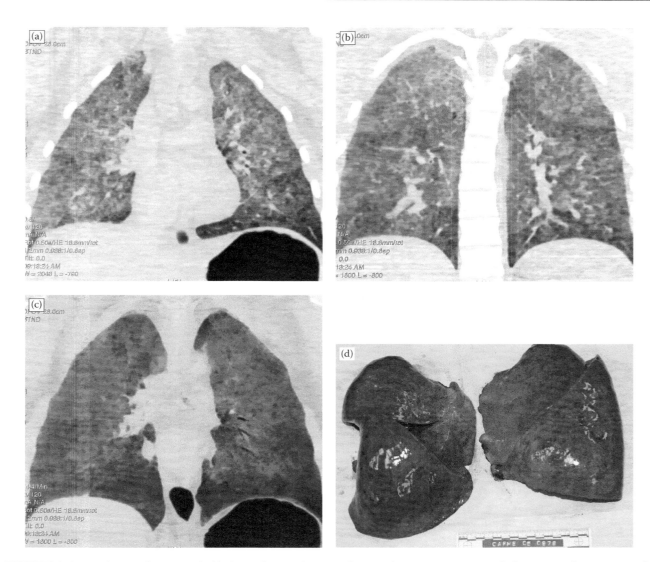

FIGURE 9.8 Freshwater drowning. (a, b) Coronal MDCT images show moderate to severe ground glass opacity heterogeneously distributed throughout the lungs. Septal lines are most prominently in the lung bases. The trachea is fluid filled. (c) Coronal minimum intensity projection shows fluid filling the trachea, main stem bronchi, and all visible airways in the right lung. A few air-filled bronchi are present on the left. The distal esophagus is distended with air, and there is air in the gastric fundus. (d) Autopsy photographs of the lungs show heavy, edematous lungs that have a dark red boggy appearance in some areas.

Imaging Findings

Pulmonary edema is the most prominent finding of drowning on plain film radiography. A variety of patterns may be seen that represent a continuum from mild to severe pulmonary edema. In mild pulmonary edema, there are interstitial and septal lines. With more severe edema, alveolar edema is present (Figure 9.4 through Figure 9.7). In reports of drowning and near drowning, pulmonary edema is usually perihilar and central in distribution. Apical and dependent patterns have also been observed in postmortem imaging. Because postmortem radiography is most often performed with the body in supine position,

fluid within the tracheobronchial tree may be difficult to observe. However, close inspection of the radiograph will often reveal the absence of the normal air-filled distal trachea and bronchi because they are fluid filled (Figure 9.5b). Radiographs should be closely inspected for high-density material in the airways and stomach which may represent inhaled or swallowed sand or silt. However, high-density sand in the airway and stomach are usually only visible on radiography when present in large amounts (Figure 9.12a,b).

MDCT closely parallels autopsy for the depiction of the anatomic findings that are supportive for the diagnosis of

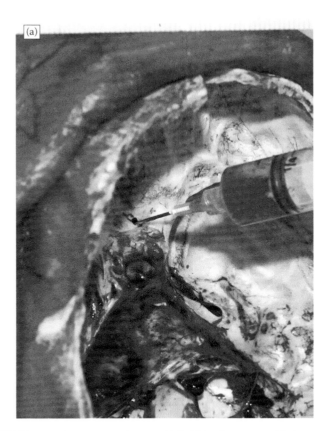 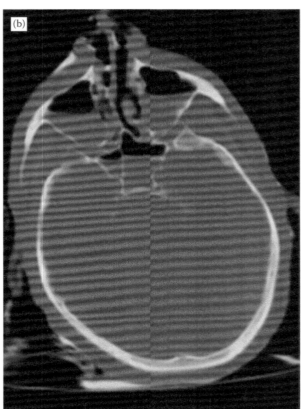

FIGURE 9.9 Sinus and mastoid fluid in drowning. (a) Autopsy photograph shows the technique of aspiration of the sphenoid sinus through the sphenoid bone. (b) Axial MDCT image shows fluid levels in both maxillary sinuses and the sphenoid sinus. A small amount of high-density sand is present on the skin of the face and in the nasal cavity.

drowning. Sinus fluid, mastoid fluid, subglottic tracheal and bronchial fluid, and pulmonary ground glass opacity are consistently present on MDCT (Levy et al. 2007). The amount of fluid within the sinuses and airways is variable. Sinuses may be completely filled with fluid, contain air fluid levels, or contain high-attenuation sand, which layers dependently in the sinus (Figures 9.9 and 9.10). Similarly, fluid or sand may be present within the trachea and bronchi (Figures 9.11 and 9.12). Minimum intensity projection images are useful to evaluate the extent of fluid-filled bronchi, because the attenuation difference between two low-attenuation structures (air in the lung and air in the bronchi) is maximized (Figures 9.3b, 9.7c, and 9.8c). Fluid is a very nonspecific finding because it may be present within the sinuses, trachea, and bronchi in other forms of death or from decomposition. However, the presence of airway froth and sand may be helpful indicators of drowning (Levy et al. 2007). Airway froth is characterized by heterogeneous low-attenuation fluid admixed with rounded foci of air (Figure 9.2b and Figure 9.3a,b). Sand, silt, or mud

appears as high-attenuation material within the sinuses or airways on MDCT (Figure 9.10 through Figure 9.12). It may also be present on the surface of the body (Figure 9.9b and Figure 9.12c,d).

Hyperinflation of the lungs is an autopsy finding indicative of drowning on opening the chest at autopsy, but there are no criteria for postmortem pulmonary hyperinflation on MDCT. However, MDCT clearly defines the severity and distribution of pulmonary edema. Several different patterns of pulmonary edema, representing a continuum of severity, may be seen in drowning deaths. Moderate ground glass opacity with septal lines and occasionally spared secondary pulmonary lobules (with a predominantly apical and perihilar distribution) is the most common pattern reported (Figure 9.4 through Figure 9.6) (Christe et al. 2008, Levy et al. 2007). Diffuse alveolar consolidation may be present in more severe pulmonary edema. Occasionally, a geographic pattern of lung attenuation characterized by areas of ground glass opacity and areas of lucency may be seen. This pattern has been called mosaic perfusion and may be secondary to focal

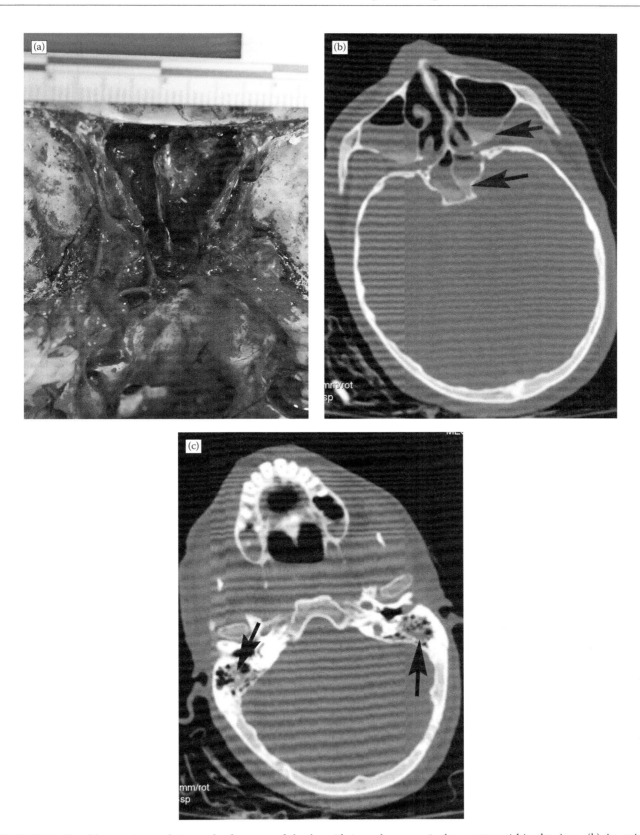

FIGURE 9.10   (a) An autopsy photograph of an unroofed sphenoid sinus shows particulate matter within the sinus. (b) An axial MDCT image shows fluid levels within both maxillary sinuses and the sphenoid sinus. There is high-density sand (arrows) in the dependent portions of each sinus. (c) Axial MDCT image shows fluid in mastoid air cells bilaterally (arrows).

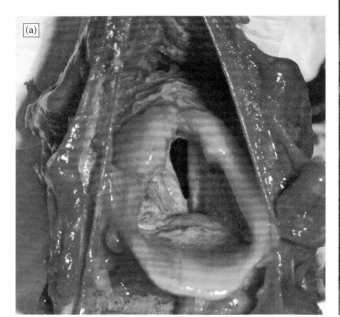

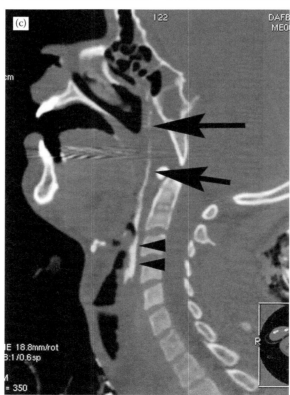

FIGURE 9.11   Sand aspiration in drowning. (a, b) Autopsy photograph of the larynx shows sand in the larynx and on the vocal cords. The opened airway in (b) shows sand below the vocal cords in the trachea. (c) Corresponding sagittal MDCT shows high-density sand throughout the pharynx (arrows) and proximal esophagus (arrowheads).

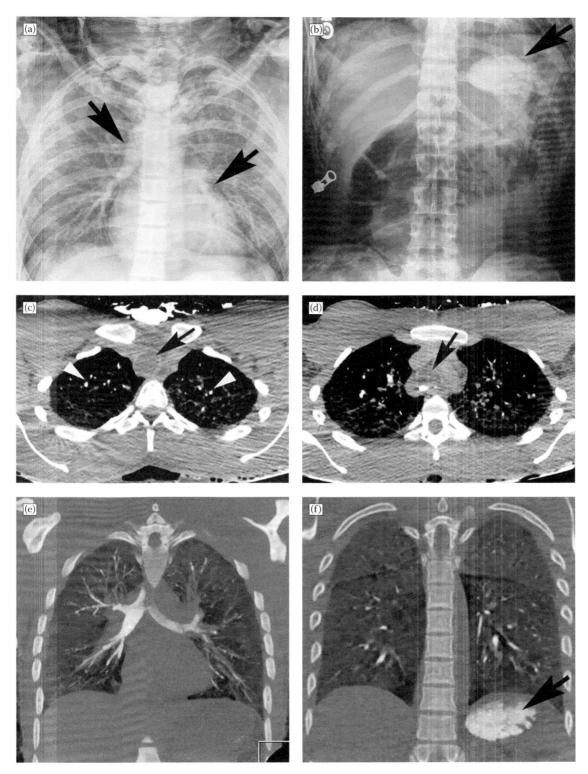

FIGURE 9.12   (a, b) Digital radiography shows high-density sand filling the tracheobronchial tree (arrows in a) and within the stomach (arrow in b). (c, d) Axial MDCT images show fluid filling the trachea (arrows) with a small amount of sand in the distal trachea (seen in d) and sand in the distal bronchi of lung apexes bilaterally (arrowheads). (e, f) Coronal MDCT images of the chest show severe pulmonary edema with diffuse alveolar opacity throughout the lungs and high-density sand filling the majority of the visualized trachea and bronchi. Ingested sand is present in the stomach (arrow in f).

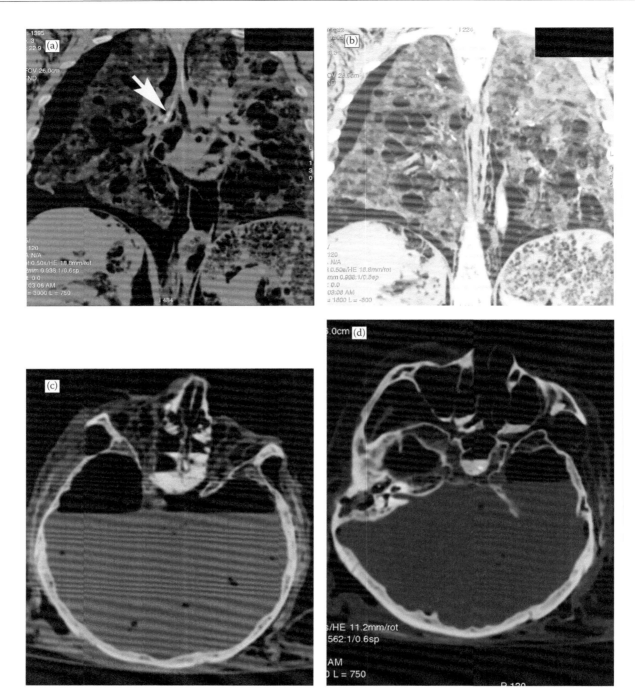

FIGURE 9.13 Sand aspiration and inhalation from drowning in a body found 2 weeks after drowning. (a, b) Coronal MDCT of the lungs shown in soft tissue (a) and lung (b) windows shows high-attenuation sand in the trachea and right main stem bronchus (arrow in a). Punctate foci of sand are present in the distal bronchi bilaterally (best shown in b). The cystic lucencies in the lung are from decomposition. Note extensive decompositional gas throughout the soft tissues. (c, d) Axial MDCT of the head shows high-attenuation sand in the ethmoid, maxillary, and sphenoid sinuses. The brain is liquefied from decomposition.

alveolar hypoxic vasoconstriction or bronchospasm adjacent to areas of alveolar edema (Figures 9.4 and 9.5) (Christe et al. 2008). Moderate and severe pulmonary edema patterns may be accompanied by pleural effusions (Figure 9.7). Dilated and engorged right-sided cardiac chambers and great vessels are nonspecific findings indicative of hypervolemia and are helpful supportive findings when other anatomic findings of drowning are present (Figures 9.14 and 9.15).

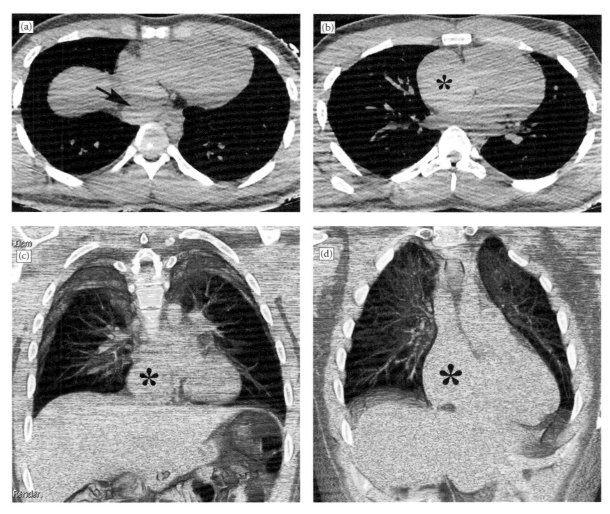

FIGURE 9.14 Right atrial enlargement in drowning. (a, b) Axial MDCT shows marked enlargement of the inferior vena cava (arrow in a) and right atrium (asterisk in b). (c, d) Color-rendered coronal maximum intensity projection right atrial enlargement (asterisks) and engorgement of the pulmonary vasculature.

Watery fluid similar to that of the drowning medium may be found in the stomach at autopsy. Sand, silt, or other debris from the drowning fluid may be present as well. The amount of fluid is variable. Consequently, the degree of fluid distension of the stomach on MDCT is not a reliable indicator that drowning fluid has been ingested at the time of death. Gaseous distension of the stomach and intestines has also been observed on MDCT and may be related to resuscitation attempts or decompositional gas (Figure 9.15).

Differential Diagnosis

Other diseases and causes of death that result in pulmonary edema and airway fluid are the principal differential diagnostic considerations from an imaging standpoint. The presence of frothy airway fluid or high-attenuation sand in the airways may be helpful findings

to suggest drowning over other causes of pulmonary edema such as cardiogenic pulmonary edema (Levy et al. 2007). Lung congestion and pleural fluid may also be seen in infectious processes or may be normal postmortem findings, especially in nontraumatic deaths. Normal postmortem lungs rarely have combined lung weights exceeding 500 to 600 grams at autopsy, and pleural effusions are usually quite small in volume, 10 to 20 mL, in normal lungs.

CONCLUSIONS

Although the diagnosis of drowning is a diagnosis of exclusion, MDCT is a helpful addition to the forensic investigation of bodies recovered from water, because the anatomic findings sought at autopsy are equally shown by MDCT. In the appropriate scenario, such as witnessed drowning,

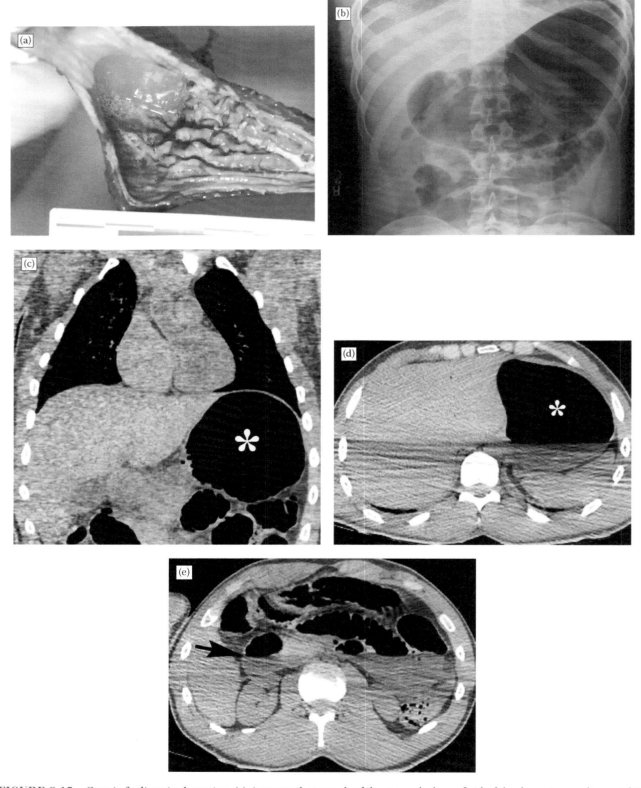

FIGURE 9.15   Gastric findings in drowning. (a) Autopsy photograph of the stomach shows fluid of the drowning medium within the stomach. (b) Anterior-posterior radiograph of the upper abdomen shows an air-distended stomach. (c, d, e) Coronal and axial MDCT images show the stomach (asterisk in c and d) and duodenum (arrow in e) distended with air and fluid. Small bowel distension is also present. Right atrial enlargement is also present on (c).

the MDCT findings of pan sinus fluid, mastoid fluid, sub-glottic tracheal and bronchial fluid, frothy fluid, sinus or airway sand, and pulmonary edema on MDCT are supportive of drowning. The main limitation of MDCT is the evaluation of vasculature. Therefore, MDCT is not helpful in excluding natural causes of death such as myocardial infarction, stroke, or pulmonary embolism. However, MDCT is a very useful adjunct to autopsy because it adds anatomic information to a cause of death rendered by external examination or limited autopsy.

REFERENCES

Azparren, J. E., Fernandez-Rodriguez, A., and Vallejo, G. 2003. Diagnosing death by drowning in fresh water using blood strontium as an indicator. *Forensic Sci Int* 137: 55–59.

Azparren, J. E., Vallejo, G., Reyes, E., Herranz, A., and Sancho, M. 1998. Study of the diagnostic value of strontium, chloride, haemoglobin and diatoms in immersion cases. *Forensic Sci Int* 91: 123–132.

Christe, A., Aghayev, E., Jackowski, C., Thali, M. J., and Vock, P. 2008. Drowning—post-mortem imaging findings by computed tomography. *Eur Radiol* 18: 283–290.

Di Maio, V. J. M., and Di Maio, D. J. 2001. *Forensic pathology*, Boca Raton, FL: CRC Press.

Gordon, I. 1972. The anatomical signs in drowning. A critical evaluation. *Forensic Sci* 1: 389–395.

Hadley, J. A., and Fowler, D. R. 2003. Organ weight effects of drowning and asphyxiation on the lungs, liver, brain, heart, kidneys, and spleen. *Forensic Sci Int* 137: 239–246.

Harcke, H. T., Levy, A. D., Abbott, R. M. et al. 2007. Autopsy radiography: digital radiographs (DR) vs. multidetector computed tomography (MDCT) in high-velocity gunshot-wound victims. *Am J Forensic Med Pathol* 28: 13–19.

Hurlimann, J., Feer, P., Elber, F. et al. 2000. Diatom detection in the diagnosis of death by drowning. *Int J Legal Med* 114: 6–14.

Idris, A. H., Berg, R. A., Bierens, J. et al. 2003. Recommended guidelines for uniform reporting of data from drowning: the "Utstein style." *Circulation* 108: 2565–2574.

Kringsholm, B., Filskov, A., and Kock, K. 1991. Autopsied cases of drowning in Denmark 1987–1989. *Forensic Sci Int* 52: 85–92.

Levy, A. D., Harcke, H. T., Getz, J. M. et al. 2007. Virtual autopsy: two- and three-dimensional multidetector CT findings in drowning with autopsy comparison. *Radiology* 243: 862–868.

Lunetta, P., Modell, J. H., and Sajantila, A. 2004. What is the incidence and significance of "dry-lungs" in bodies found in water? *Am J Forensic Med Pathol* 25: 291–301.

Modell, J. H. 1978. Biology of drowning. *Annu Rev Med* 29: 1–8.

Modell, J. H. 1993. Drowning. *N Engl J Med* 328: 253–256.

Modell, J. H., Bellefleur, M., and Davis, J. H. 1999. Drowning without aspiration: is this an appropriate diagnosis? *J Forensic Sci* 44: 1119–1123.

Modell, J. H., Gaub, M., Moya, F., Vestal, B., and Swarz, H. 1966. Physiologic effects of near drowning with chlorinated fresh water, distilled water and isotonic saline. *Anesthesiology* 27: 33–41.

Modell, J. H., Moya, F., Newby, E. J., Ruiz, B. C., and Showers, A. V. 1967. The effects of fluid volume in seawater drowning. *Ann Intern Med* 67: 68–80.

Pearn, J. 1985. Pathophysiology of drowning. *Med J Aust* 142: 586–588.

Chapter 10

Suicide

FORENSIC PRINCIPLES

The manner of death is deemed suicide when a person intentionally kills himself or herself. Suicide is receiving increased attention worldwide because many countries are developing and instituting national strategies for prevention (Mann et al. 2005). Suicide rates vary from country to country and by age, gender, ethnic origin, employment status, and occupation. In general, suicide is more common in men than women. Most experts believe that suicide is multifactorial and not likely the consequence of a single cause or stressor. Psychosocial and psychiatric factors are classified as proximal factors or triggers to suicide. Familial or genetic tendencies, childhood experiences, and underlying metabolic conditions are classified as distal or predisposing factors (Hawton and van Heeringen 2009). The psychological autopsy is the traditional method of studying the characteristics of individuals who die by suicide. This approach supports a strong link between suicide and psychiatric conditions such as affective disorders, substance abuse, and schizophrenia (Cavanagh et al. 2003). In 2006, suicide was the eleventh leading cause of death for all ages in the United States. Males accounted for 79% of the cases, four times the rate of female suicide. Suicide was the second leading cause of death among 25- to 34-year-olds and the third leading cause of death among 15- to 24-year-olds (Centers for Disease Control and Prevention 2008).

When a person is contemplating suicide, access to specific methods may be a factor that leads that person from suicidal thoughts to suicidal action (Ajdacic-Gross et al. 2008). Suicidal methods vary in the likelihood that an attempt will be fatal (Hawton and van Heeringen 2009). Rich divides methods of suicide into two major groups: those that are immediately fatal (firearms, explosive devices, hanging, jumping, self-immolation, electrocution) and those that are less likely to be immediately fatal (overdose, poisoning, drowning, cutting, and stabbing) (Rich et al. 1998). In the United States, firearms, drug overdose, and hanging are the most common methods used. Self-inflicted sharp force injury, jumping, and intentionally crashing a vehicle, while not common methods of suicide, are known to occur. In other parts of the world,

self-immolation and suicide bombing are seen more commonly. The victim in 10% to 50% of cases, depending on the study cited, leaves a suicide note. For general reference, a note is left in 25% to 30% of cases (Dolinak et al. 2005, Koehler 2007, Spitz et al. 2006).

Firearms are used in approximately 60% of all suicide deaths in the United States. The use of firearms in suicide varies among localities within the United States because of regional differences in firearm availability and law enforcement (Brent 2001). Di Maio and Di Maio note that even though firearms are traditionally a preferred method for men to commit suicide, their experience shows that firearms are becoming a method of suicide preferred by women as well (Di Maio and Di Maio 2001). Handguns are used more commonly than rifles or shotguns, especially with female deaths. In firearms deaths, there may be an attempt to make the suicide seem accidental, and in homicides, an attempt may be made to make the death appear suicidal. The circumstances of death, findings at the scene of death, weapon characteristics, toxicology, and autopsy findings are used in cases to make the final determination of the cause and manner of death.

Hanging is one of the most commonly used methods of suicide worldwide (Gunnell et al. 2005). It has been the second or third most popular method in the United States depending on area of the country and gender of the victim. In hanging, death occurs from asphyxia because the ligature or noose that surrounds the neck compresses the carotid arteries, reducing cerebral blood flow (Spitz et al. 2006). The victim's body may be partially or completely suspended in suicidal hanging. Victims can hang themselves in the sitting, kneeling, or lying down positions. In the majority of cases, the noose is constructed of a single loop of any material that is handy. Ropes, electric cords, and belts are commonly used. In jails and prisons, sheets can be torn into strips and items of clothing can be fashioned into a noose (Di Maio and Di Maio 2001).

Toxicology plays an important role in the conduct and investigation of suicide. Drug overdose vies with hanging as the second or third most common method of suicide in the United States. The drug of choice changed over the past two decades from barbiturates to tricyclic antidepressants (Di Maio and Di Maio 2001). The importance of toxicology results is obvious in establishing overdose as the cause of death, but psychoactive substances are also recognized as suicide-generating stimuli, so the presence of nonlethal amounts of drugs is important. Alcohol is known to produce disinhibition and increased self-confidence. Drugs that cause central nervous system stimulation can create a degree of enhanced aggression (Coklo et al. 2009). The presence of these substances can contribute to suicidal tendencies. Even though pesticides are commonly used for suicide in some parts of the world such as Asia and Latin America, they are usually not used in the United States and Europe (Ajdacic-Gross et al. 2008).

Death in jumping and vehicular suicide is from blunt force injury. History, circumstances of death, witnesses, scene investigation, and toxicology are important to establish the manner of death in these cases. It is also possible for more than one method to be involved in a suicide. For example, if drug ingestion does not work fast enough, it could be followed by gunshot or hanging. Suicidal fire deaths are rare. Self-immolation is a suicidal method in some cultures and may be performed during an acute emotional reaction, by a person with a fire fetish, or for idealistic purposes (Dolinak et al. 2005).

The forensic pathologist must distinguish suicide from other manners of death. In some instances, it may be difficult. Spitz discusses example cases of asphyxia where the victim was found with extremities bound and connected in some manner to a noose around the neck. These cases aroused suspicion of homicide or suicide but were often associated with sexual acts and ultimately concluded to be accidental (Spitz et al. 2006).

Autopsy Findings

In suicide by gunshot wound, the location of self-inflicted wounds depends on weapon used, gender of the victim, and to some extent, right- or left-handedness. Individuals who shoot themselves with handguns choose the head (81%), chest (17%), and abdomen (2%) as anatomic locations to kill themselves. The same pattern holds for suicides with rifles and shotguns: head (69%), chest (28%), and abdomen (3%) (Di Maio 1999). Most suicidal gunshot wounds are hard or loose contact wounds; however, a small number are intermediate range wounds (Di Maio 1999). Contact wounds can show muzzle imprint and soot deposit on the skin (see Chapter 4, Figure 4.5). The weapon may be held and fired in an atypical position. For instance, the butt of the weapon can be held in the hand with the thumb on the trigger. Often the barrel of the gun will be held in the non-firing hand, which can produce powder tattooing from blowback of powder exiting the muzzle (see Chapter 4, Figure 4.6). Multiple self-inflicted gunshot wounds may be found if the wounds are not immediately incapacitating. Gunshot wounds entering the brain and cervical spinal cord are immediately incapacitating, whereas gunshots to the heart and aorta may not be immediately fatal. Death may take 10 to 15 seconds, which is enough time for the victim to be capable of potential movement (Dolinak et al. 2005). In addition, multiple gunshots can be self-inflicted if an automatic weapon is used (Figure 10.1).

Suicidal hanging may occur with or without marks on the neck. The presence of a groove or furrow on the neck can suggest the type of noose or loop used. Horizontal skin folds and creases on the neck of obese persons or infants may resemble a noose mark. Genuine noose marks remain after the bloating and skin slippage of decomposition. Over 50% of suicidal hangings show petechial congestion hemorrhages on the face and neck above the level of the noose. These may also be present in the mouth and conjunctiva (see Chapter 12, Figure 12.1). Injury to the strap muscles is rare in hanging, and fracture of the thyroid cartilage or hyoid bone is the exception rather than the rule. The victim may have nail marks and bruises on the neck and fingers secondary to a reflex action to preserve life (Spitz et al. 2006).

It is unusual for the body to drop a long distance in suicidal hangings. If it does drop a long distance before being caught by a taut rope, the classic cervical vertebral injury of judicial hanging may occur, and in extreme cases, decapitation is possible. In these hangings, fracture of the neural arch of the C2 vertebrae with fracture-dislocation of C2 from C3 occurs with stretching or tearing of the cervical spinal cord (Figure 10.2) (Dolinak et al. 2005).

People who commit suicide by sharp force injury often have multiple incised wounds of variable depths on the neck, antecubital fossa, or flexor aspects of the forearms and wrists (Dolinak et al. 2005). Self-inflicted stab or incised wounds are analyzed in context with historical information, terminal events, and the scene findings.

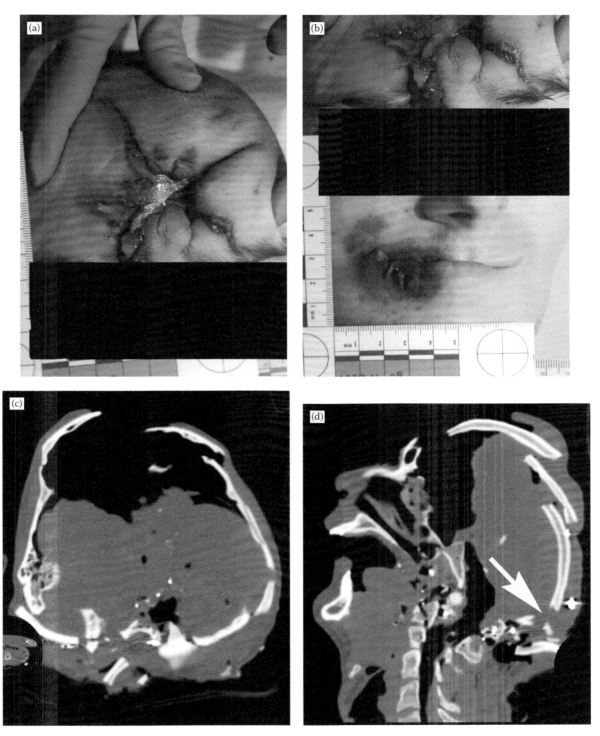

FIGURE 10.1 Suicide with multiple semiautomatic gunshot wounds to the head. (a) Autopsy photograph of the forehead shows a stellate entrance wound with soot in the pattern produced by the flash suppressor on the muzzle of a rifle. (b) Autopsy photograph of a second contact wound at the margin of the mouth also shows soot. Two wound tracks can be identified on MDCT. (c, d) Axial oblique and sagittal oblique images show a wound track entering the frontal bone passing through the brain and exiting in the occiput (arrow in d).

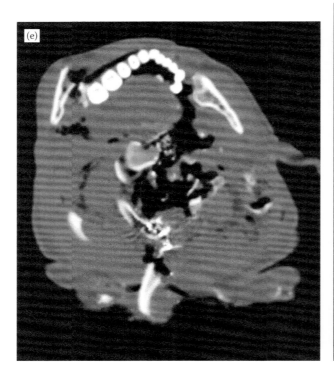

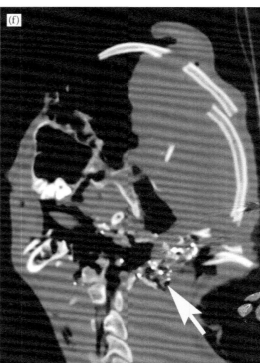

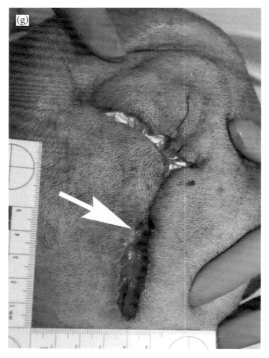

FIGURE 10.1 (*Continued*) Suicide with multiple semiautomatic gunshot wounds to the head. (e, f) Axial oblique and sagittal oblique images show a second track passing through the oral cavity and upper cervical spine and exiting below the occiput (arrow in f). Small metal fragments are in the track. (g) Autopsy photograph shows two exit wounds on the back of the head, an upper stellate wound, and a lower elongated wound (arrow). Based on autopsy and scene investigation, the forensic pathologist concluded that the head wound occurred first and the facial wound second as the semiautomatic weapon shifted position.

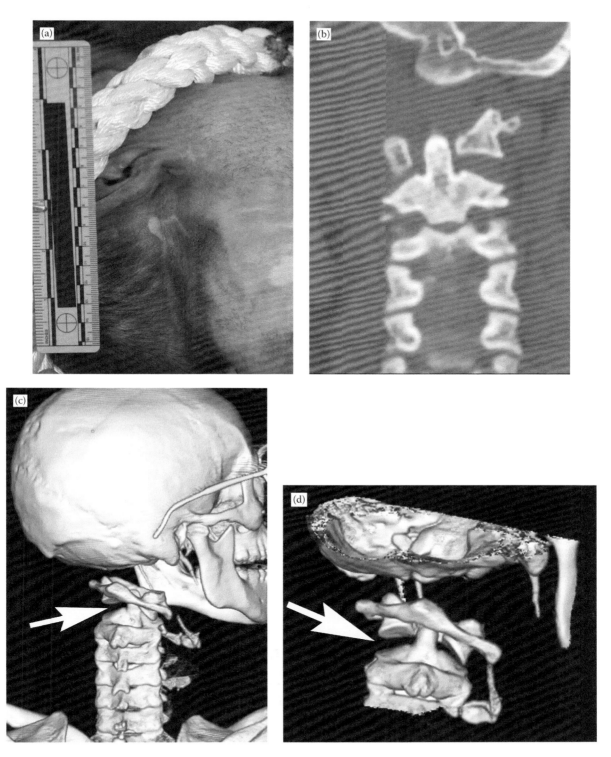

FIGURE 10.2   Suicide by hanging. The victim's body dropped a long distance before being caught by a taut rope, resulting in a judicial-type hanging with disruption of the upper cervical spine. (a) Autopsy photograph shows a broad, reddish brown furrow on the victim's neck matching the adjacent rope. (b, c) Coronal and three-dimensional MDCT show occipito-atlanto and atlanto-axial (C1-C2) dissociation (arrow). (d, e) Three-dimensional computed tomography following digital subtraction of bone shows details of the C1-C2 dissociation (arrow in d).

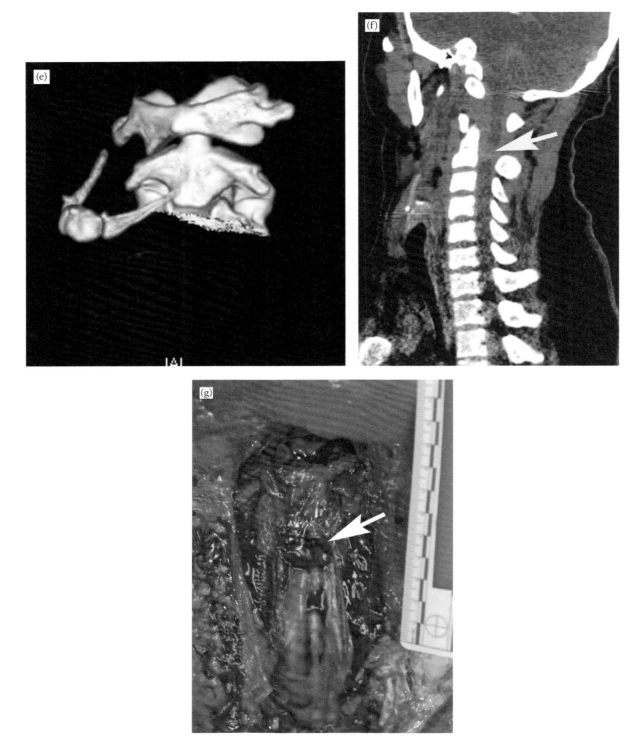

FIGURE 10.2 (*Continued*) Suicide by hanging. The victim's body dropped a long distance before being caught by a taut rope, resulting in a judicial-type hanging with disruption of the upper cervical spine. (d, e) Three-dimensional computed tomography following digital subtraction of bone shows details of the C1-C2 dissociation (arrow in d). (f) Sagittal MDCT shown in soft tissue window shows a focus of high-attenuation hemorrhage and cord discontinuity behind the body of C2 (arrow). (g) Autopsy photograph shows spinal cord transection with a focal hemorrhage at the site of transection (arrow).

Establishing poisoning and drug overdose as a cause of death relies on toxicology. The autopsy is necessary to exclude other causes of death and to look for contributory factors. Anatomic findings such as scars on the wrist may indicate that there were previous suicide attempts, or in the case of carbon monoxide toxicity, the finding of cherry red coloration in soft tissues supports carbon monoxide inhalation.

In assessing suicide death by blunt force injury, the pattern of injury is assessed in context with the details from the scene of death. Falling bodies can strike protruding structures or bounce on impact, producing additional injury. Circumstances at the scene of a vehicular crash suicide must fit the injuries documented at autopsy.

RADIOLOGIC PRINCIPLES

Goals of Imaging

The goals of imaging in cases of suicide depend on the method of suicide. For example, in gunshot wound deaths, imaging seeks to determine entry and exit wounds, bullet trajectory, tissue damage, and to assist with recovery of ballistic material and evidence. In contrast, the goal of imaging in drug overdose is to reveal unsuspected findings that might indicate a different cause of death. Determination of manner of death is the task of the forensic pathologist, but the radiologist should seek clues that suggest one manner of death may be more likely than another.

Imaging Findings

In suicide by firearm, the imaging findings parallel those of gunshot wounds in general, but there are patterns that are more consistent with suicide. There is usually a single gunshot wound in suicides, but more than one wound does not exclude suicide. Multiple gunshot wounds may involve more than one area of the body and are possible when individual shots are not lethal. Multiple wounds can also be present when a semiautomatic or automatic weapon is used (Figure 10.1).

Contact gunshot wounds to the head predominantly involve five entry sites: frontal or glabella, right temporal (see Chapter 4, Figure 4.5), left temporal, intraoral (Figure 10.3), and submental or submandibular (Figure 10.4). If the bullet perforates the skull, the exit wound is on the opposite side of the skull for temporal gunshots and generally in the posterior or occipital region for frontal and intraoral

gunshots. The exit site for submental gunshots varies with the angle of the gun. Some exit in the high frontal area and others in the vertex or high occiput. The cavitation produced by contact gunshot wounds may produce an entrance wound that is larger than the exit wound. This is commonly observed in intraoral and submental tissues. We studied three-dimensional surface-rendered MDCT and found that the image detail of the wound on MDCT lacks the clarity needed for forensic conclusions. Another limitation of MDCT is its inability to show skin coloration, soot, and powder tattooing from muzzle blowback.

MDCT is very accurate in documenting entry and exit wounds as well as the wound track in gunshot wounds to the head (Levy et al. 2006). Beveling of the skull wound margin is used to determine direction of the bullet. Internal beveling is present at the entry wound and external beveling at the exit wound (Harcke et al. 2008). Bone fragments travel in the direction of the projectile. Multiplanar and three-dimensional MDCT should be used to assess bone and soft tissue findings that determine the wound track (Figure 10.1 through Figure 10.3). Because of the differences between the body position at the time of shooting and the position of the body when the postmortem MDCT is acquired, bone and soft tissue findings will usually be nonlinear with respect to the expected path of the bullet. The position of organs may also change from the time of death to postmortem examination. For example, gunshot wounds to the chest may collapse the lung and cause positional shifts of the diaphragm and heart, which make the gunshot wound track appear nonlinear on MDCT images (Figure 10.5).

The furrow or ligature mark seen in some cases of suicide by hanging may be difficult to establish on MDCT. It must be differentiated from skin folds and creases. This can be especially difficult in obese victims. Similarly, surface marks made by a knot at the point of suspension may be obscure. Skin surface impressions from a noose are best depicted by three-dimensional surface-rendered MDCT (see Chapter 12, Figure 12.2). We must note, however, that the furrow mark is not always visible by MDCT, and the coloration typical of furrow marks cannot be evaluated on MDCT (see Chapter 12, Figure 12.1). In the uncommon situation where a long drop occurs in hanging, MDCT is very useful to evaluate high cervical fractures and dislocations and thyroid cartilage and hyoid fractures. In Figure 10.2, a suicide with long drop hanging, the cervical spine is separated

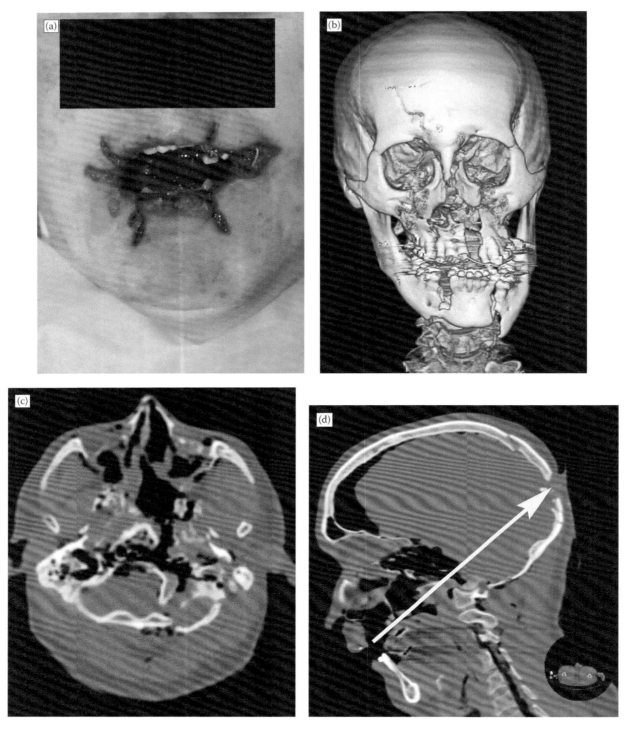

FIGURE 10.3 Suicide by intraoral gunshot. (a) Autopsy photograph of the mouth shows stellate perioral laceration with intraoral soot deposition in the entry wound. (b) Three-dimensional MDCT shows complex fractures of the mandible, maxilla, and calvarium. (c, d) Axial and sagittal oblique MDCT in the plane of the wound track (arrow in d) show that the bullet passed through the base of the skull and exited through the occiput.

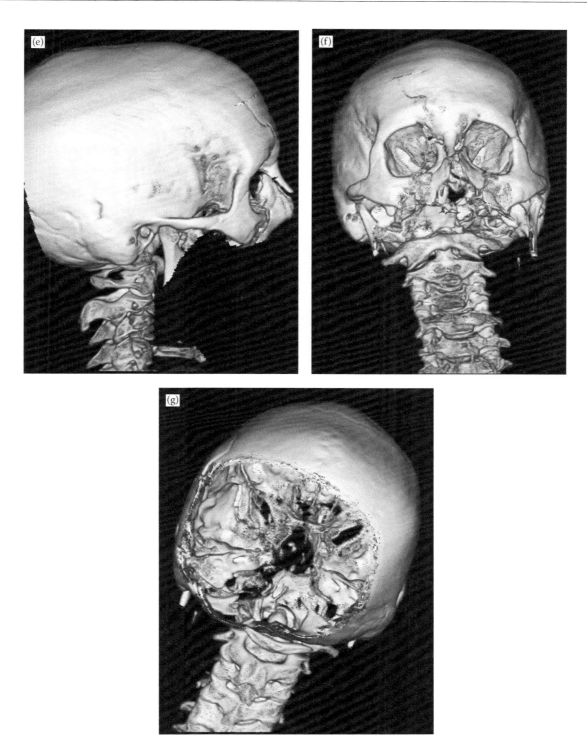

FIGURE 10.3   (*Continued*) Suicide by intraoral gunshot. (e, f, g) Digital dissection using three-dimensional MDCT processing shows removal of facial structures and occiput to display fracture defects in the skull base as viewed from anterior (f) and posterior (g).

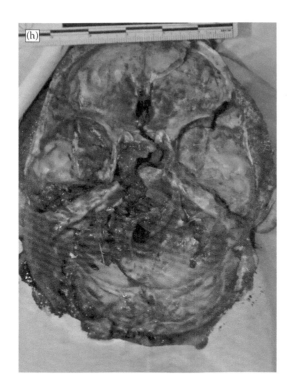

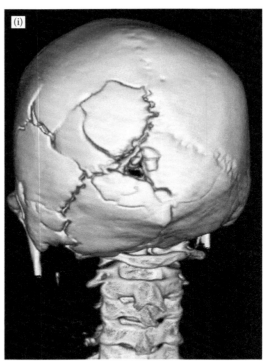

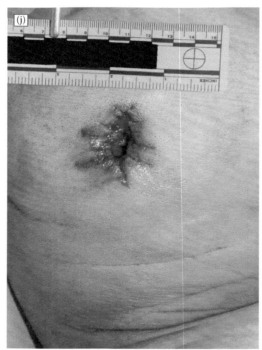

FIGURE 10.3   (*Continued*) Suicide by intraoral gunshot. (h) Autopsy photograph of the skull base correlates with (f) and (g) and shows hemorrhage and comminuted fractures throughout the entire skull base. (i) Three-dimensional MDCT of the posterior skull shows the occipital exit defect with associated fractures. (j) Autopsy photograph of the back of the head shows the occipital exit wound.

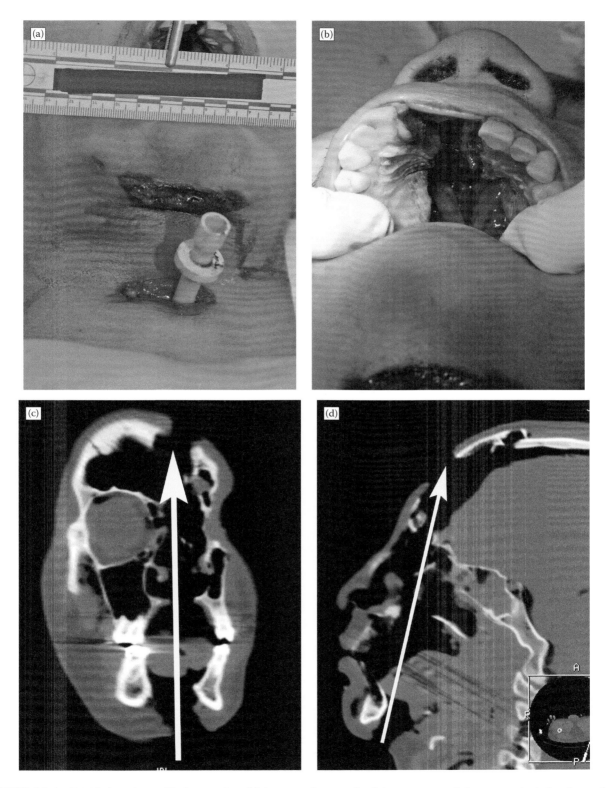

FIGURE 10.4 Suicide by submandibular gunshot. (a) Autopsy photograph of the entry wound shows marginal abrasion and soot deposition. The tracheostomy tube is from attempted resuscitation. (b) Autopsy photograph shows that the bullet passed through the hard palate, fracturing the maxilla. (c, d) Coronal and sagittal oblique MDCT show the wound track (arrow shows bullet path) passed through the nasal cavity and exited in the frontal calvarium.

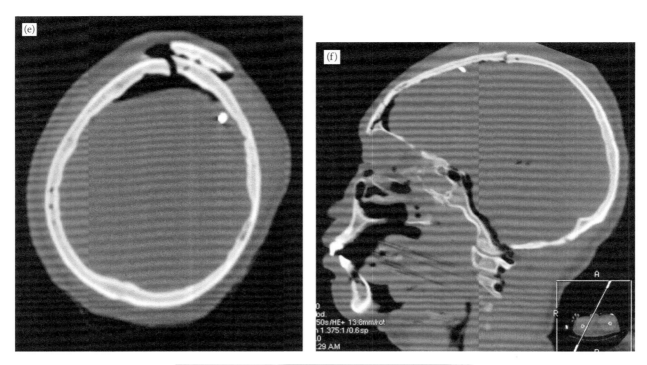

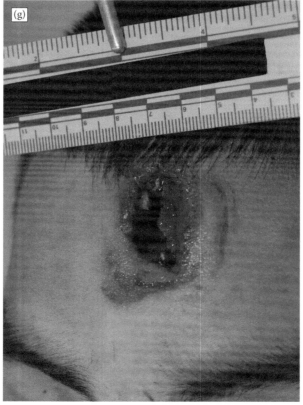

FIGURE 10.4 (*Continued*) Suicide by submandibular gunshot. (e, f) Axial and sagittal MDCT show the frontal bone fracture and a bullet fragment in the left frontal region of the brain. (g) Autopsy photograph shows the frontal exit wound.

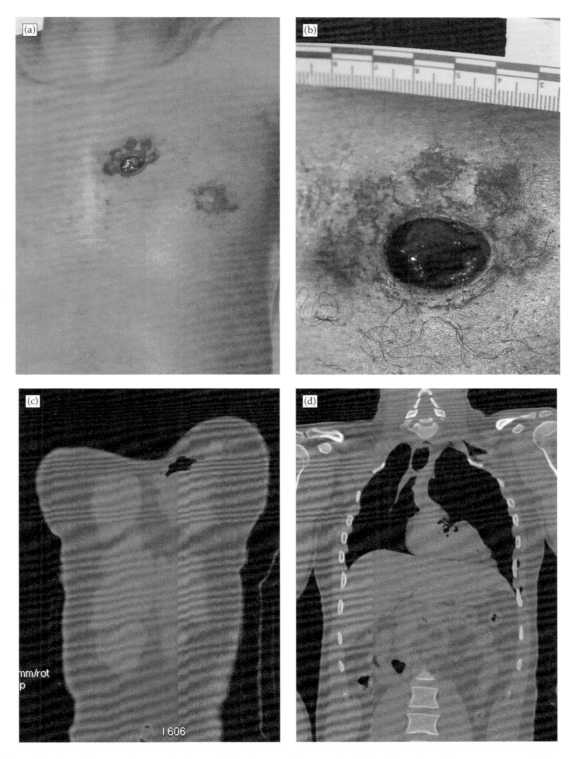

FIGURE 10.5   Suicide by gunshot to the chest. (a, b): Autopsy photographs of the chest show the entrance wound has soot and a distinctive contact pattern typical of the flash suppressor on the muzzle of a rifle. (c, d, e, f) Serial coronal MDCT images reflecting the wound track: (c) anterior left chest wall, (d) upper heart, (e) ninth posterior left rib, and (f) exit on the back.

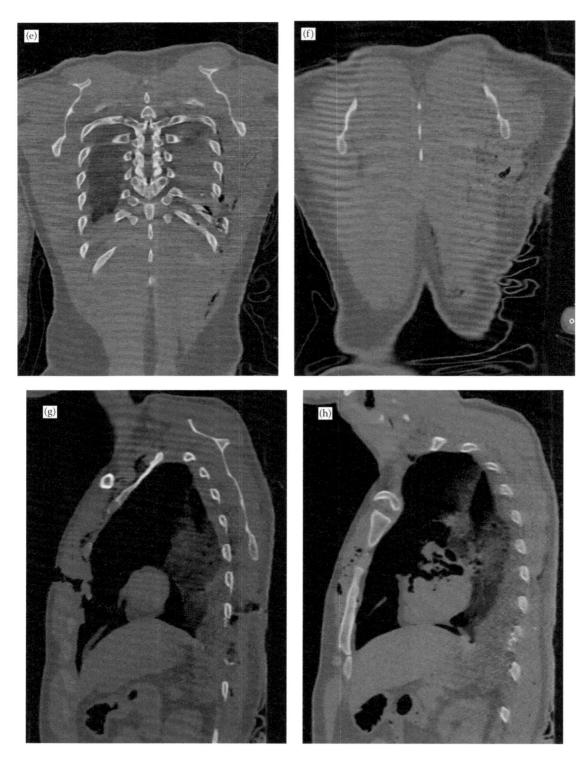

FIGURE 10.5 (*Continued*) Suicide by gunshot to the chest. (c, d, e, f) Serial coronal MDCT images reflecting the wound track: (c) anterior left chest wall, (d) upper heart, (e) ninth posterior left rib, and (f) exit on the back. (g, h) Sagittal oblique MDCT in the plane of the wound track shows the entry and exit wounds in (g) and the heart defect and ninth posterior left rib fracture in (f). The bullet track is nonlinear because the lung collapsed and the victim is in the supine position for the computed tomography scan.

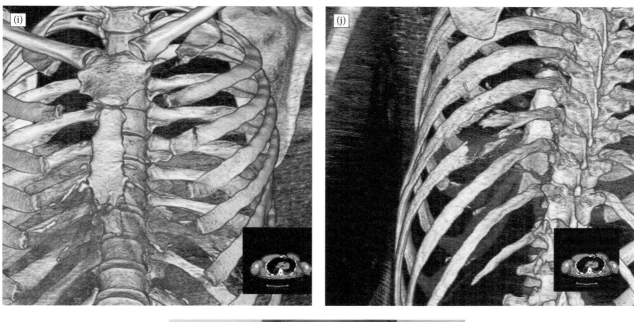

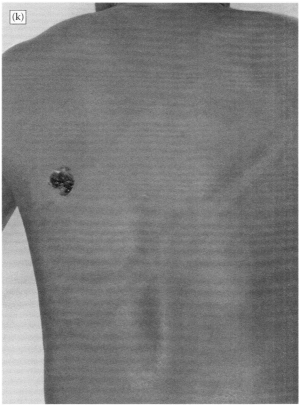

FIGURE 10.5 (*Continued*) Suicide by gunshot to the chest. (i, j) Three-dimensional MDCT shows fractures of the anterior third left rib and ninth posterior left rib. (k) Autopsy photograph of the left back shows the exit wound below the left scapula.

at both the occipito-atlanto and atlanto-axial (C1-C2) levels with spinal cord transection. In some suicide hanging cases, postmortem MDCT revealed pneumomediastinum and subcutaneous emphysema of the neck below the level of the strangulation mark. This was described as a sign of vitality at the time of hanging (Aghayev et al. 2005). Care must be taken to differentiate this finding from decompositional gas.

The mechanism of death in drug overdose is cardiorespiratory death, and as such, the findings on MDCT are nonspecific. Gastric contents are evaluated in every autopsy and should also be checked on MDCT. Round or oval high-attenuation dense material may reflect ingestion of pills. Interpretation of gastric contents should be made with caution, because it is well known that digestion continues in the stomach for a considerable amount of time after death (Spitz et al. 2006). Suicide by jumping or vehicle crash is studied as blunt force injury. Careful documentation of injury pattern may help the forensic pathologist reconstruct the details of the incident.

CONCLUSIONS

Determination of suicide as the manner of death is made by the forensic pathologist using data collected by autopsy, toxicology, scene investigation, history, and witnesses. MDCT adds supporting data and is most useful in firearm deaths, where it can provide specific information. In contrast, MDCT is complementary and helps to exclude unsuspected pathology or injury in hanging and drug-related deaths.

REFERENCES

Aghayev, E., Yen, K., Sonnenschein, M. et al. 2005. Pneumomediastinum and soft tissue emphysema of the neck in postmortem CT and MRI; a new vital sign in hanging? *Forensic Sci Int* 153: 181–188.

Ajdacic-Gross, V., Weiss, M. G., Ring, M. et al. 2008. Methods of suicide: international suicide patterns derived from the WHO mortality database. *Bull World Health Organ* 86: 726–732.

Brent, D. A. 2001. Firearms and suicide. *Ann N Y Acad Sci* 932: 225–239; discussion, 239–240.

Cavanagh, J. T., Carson, A. J., Sharpe, M., and Lawrie, S. M. 2003. Psychological autopsy studies of suicide: a systematic review. *Psychol Med* 33: 395–405.

Centers for Disease Control and Prevention. 2008. National center for health statistics. http://www.cdc.gov/nchs/fastats/deaths.htm (accessed October 16, 2008).

Coklo, M., Stemberga, V., Cuculic, D., Sosa, I., and Bosnar, A. 2009. Toxicology and methods of committing suicide other than overdose. *Med Hypotheses* 73: 809–810.

Di Maio, V. J. M. 1999. *Gunshot wounds: practical aspects of firearms, ballistics, and forensic techniques,* Boca Raton, FL: CRC Press.

Di Maio, V. J. M., and Di Maio, D. J. 2001. *Forensic pathology,* Boca Raton, FL: CRC Press.

Dolinak, D., Matshes, E., and Lew, E. O. 2005. *Forensic pathology: principles and practice,* Burlington: Elsevier Academic Press.

Gunnell, D., Bennewith, O., Hawton, K., Simkin, S., and Kapur, N. 2005. The epidemiology and prevention of suicide by hanging: a systematic review. *Int J Epidemiol* 34: 433–442.

Harcke, H. T., Levy, A. D., Getz, J. M., and Robinson, S. R. 2008. MDCT analysis of projectile injury in forensic investigation. *AJR Am J Roentgenol* 190: W106–W111.

Hawton, K., and van Heeringen, K. 2009. Suicide. *Lancet* 373: 1372–1381.

Koehler, S. A. 2007. The role of suicide notes in death investigation. *J Forensic Nurs* 3: 87–88, 92.

Levy, A. D., Abbott, R. M., Mallak, C. T. et al. 2006. Virtual autopsy: preliminary experience in high-velocity gunshot wound victims. *Radiology* 240: 522–528.

Mann, J. J., Apter, A., Bertolote, J. et al. 2005. Suicide prevention strategies: a systematic review. *JAMA* 294: 2064–2074.

Rich, C. L., Dhossche, D. M., Ghani, S., and Isacsson, G. 1998. Suicide methods and presence of intoxicating abusable substances: some clinical and public health implications. *Ann Clin Psychiatry* 10: 169–175.

Spitz, W. U., Spitz, D. J., and Fisher, R. S. 2006. *Spitz and Fisher's medicolegal investigation of death: guidelines for the application of pathology to crime investigation,* Springfield, IL: Charles C Thomas.

Chapter 11

Natural Death

FORENSIC PRINCIPLES

Natural death is death from a naturally occurring disease process. The most common cause of natural death in the United States is heart disease. Cancer, cerebrovascular disease, and respiratory diseases are other leading causes of natural death (Centers for Disease Control and Prevention 2008). Sudden death is a subset of natural death that is usually described as natural unexpected death occurring within an hour after the onset of symptoms. The most common cause of sudden death is cardiovascular disease, namely atherosclerotic coronary artery disease. Other forms of cardiovascular disease such as hypertrophic cardiomyopathy, right ventricular cardiomyopathy, myocarditis, coronary artery anomalies, coronary artery dissection, aortic dissection, and conductive system abnormalities may also cause sudden death. Noncardiac causes of sudden death include sudden unexplained death in epilepsy, cerebrovascular disease, asthma, pulmonary embolism, and intra-abdominal hemorrhage.

In many natural deaths, medical examiners do not investigate or perform autopsy because death is expected. This is especially true in elderly individuals or individuals with known diseases. Therefore, the majority of natural deaths investigated by medical examiners are those that occur suddenly or unexpectedly, especially in younger individuals. For most medical examiners, the term *younger* refers to individuals under the age of 50. In the course of the autopsy, unnatural causes such as violent and toxic causes of death should always be excluded. Medical examiners begin each autopsy with a level of suspicion that a homicide has occurred and then develop the cause and manner of death from that perspective. The cause of death cannot be determined in 5% of cases after a complete postmortem and forensic examination (de la Grandmaison and Durigon 2002).

AUTOPSY FINDINGS

The forensic autopsy of a victim of apparent natural or sudden death requires information from the scene investigation and interviews with family members, witnesses, and emergency responders when applicable. The victim's past medical history and family history may provide valuable information about the victim's current medical condition and risk factors for sudden death. In addition to a detailed external examination and autopsy dissection, serum biochemical analysis, fluid analysis, microbiology cultures, and toxicology often provide useful clues as to the cause of death. Specifically, toxicology may establish that drugs or toxins are the cause of death or may help to determine if there are drugs present that may have contributed to sudden death or may be the cause of disorders such as drug-related cardiomyopathies (de la Grandmaison 2006).

Traumatic injuries may be evident in natural deaths if the individual experienced a loss of consciousness that resulted in a more violent circumstance surrounding death, such as a fall with significant blunt traumatic injuries, an automobile accident, or drowning. In some instances, it may be difficult to discern these injuries from inflicted trauma. The presence or absence of associated bruising and hemorrhage may help to determine if trauma occurred antemortem or postmortem. Resuscitative injuries should be distinguished from traumatic injuries. Needle punctures, defibrillator marks, anterior rib and sternal fractures, and neck injuries are commonly present. Rib and sternal fractures associated with resuscitation are typically seen in adults and generally have minimal associated hemorrhage. Occasionally, intubation injuries to the cricopharyngeal region, vallecula, and piriform sinuses may simulate strangulation injuries.

Specific anatomic areas are very carefully evaluated for abnormalities that commonly cause sudden or natural death. Atherosclerotic and thrombotic findings in coronary, cerebral, and pulmonary vasculature can confirm cardiovascular disease as the cause of death. The left ventricular papillary muscle, subendocardial myocardium and interventricular septum, cerebellum, and basal ganglia may show evidence of ischemia as the cause of death (Christiansen and Collins 2007). When necessary, the cardiac conduction system may be retained for later evaluation if arrhythmias are of concern (Hutchins 2003). The lungs are sectioned for evidence of pulmonary disease, and the larynx is evaluated for evidence of edema that may have occurred in an anaphylactic reaction. The remainder of the visceral organs are examined and sectioned looking for evidence of malignancy and other natural diseases.

Cardiovascular Disease

Cardiovascular disease is the most common cause of natural sudden death. Decedents may or may not have experienced the classic symptoms of myocardial infarction, such as angina or chest pain prior to death and may not have had a prior history of cardiac disease. The mechanism of death in atherosclerotic coronary artery disease is a fatal arrhythmia, such as a ventricular tachyarrhythmia or asystole. The most common finding at autopsy is eccentric atherosclerotic luminal narrowing of greater than 75% in one or more coronary arteries (Figures 11.1 and 11.2). Two-vessel disease is usually found, but a single vessel, namely the proximal left anterior descending artery, with severe disease may also be fatal. Myocardial scarring is the second most common finding (Di Maio and Di Maio 2001). Acute infarction may have visible areas of edema, myocardial necrosis, or hemorrhage at autopsy (Figure 11.3). If dating of the infarction is necessary, histological evaluation of the heart is performed. Evidence of prior chronic cardiac disease and intervention may be present. Pacemakers, coronary artery stents (Figure 11.4), and other devices should always be noted. Hypertensive cardiovascular disease is characterized by cardiomegaly and left ventricular hypertrophy. It is usually accompanied by atherosclerosis that produces concentric narrowing of the coronary arteries. Other cardiac disorders associated with sudden death include bridging of the left anterior descending artery, cardiomyopathies, myocarditis, and valvular diseases.

Aortic aneurysm and aortic dissection are natural diseases of the aorta that may also result in sudden death. Aortic dissection occurs when blood enters the wall of the aorta through a tear in the aortic intima and separates the layers of the media, creating a false channel (Figures 11.5 and 11.6). The diagnosis of aortic dissection is based on finding an intimal tear, which usually runs transversely or longitudinally with respect to the length of the aorta. The tear leads to an intramural hematoma. In fatal cases, aortic dissection extends into the coronary arteries or leads to a pericardial hematoma. Although the majority of decedents with aortic aneurysms and dissection have a history of hypertension and atherosclerotic vascular disease, underlying diseases and genetic disorders should be excluded, because diseases of the aorta may be secondary to Marfan syndrome, Ehlers-Danlos, cystic medial necrosis, or long-term cocaine use (Di Maio and Di Maio 2001).

Intracranial and Cerebrovascular Disease

Epilepsy is the most common intracranial cause of sudden death. Spontaneous subarachnoid hemorrhage, intracerebral hemorrhage, tumors, and meningitis account for the remainder of intracranial causes. Sudden unexpected death in epilepsy is defined as sudden death in a person with epilepsy with or without seizure activity in whom there is no demonstrable cause of death after complete autopsy, toxicology, and scene investigation (Shields et al. 2002). The mechanism of death is not fully understood, and the findings at autopsy are not consistent. The most common finding at autopsy is bite marks or contusions on the tongue from seizure activity. Pulmonary edema may be present. Atrophy from prior injury may be found in the cerebral cortex, hippocampus, or cerebellum. Although a history of seizure activity is not present in all cases, a well-documented clinical history of epilepsy or seizure activity in the correct setting is helpful to establish the cause of death when no conclusive findings are present at autopsy.

Spontaneous, nontraumatic subarachnoid hemorrhage is the second most common cause of intracranial sudden death. Subarachnoid hemorrhage most commonly occurs from rupture of a cerebral aneurysm, which is often referred to as a *berry aneurysm*. Intracerebral hemorrhage, arteriovenous malformations, blood dyscrasias, anticoagulation, tumors, mycotic aneurysms, and vasculitis are other causes of spontaneous subarachnoid hemorrhage. Subarachnoid hemorrhage from cerebral aneurysm rupture may extend into the brain parenchyma or intraventricular spaces. Rarely, hemorrhage from cerebral aneurysms will be isolated to the brain parenchyma or ventricles with no evidence of subarachnoid blood (Gonsoulin et al. 2002). Although cerebral aneurysms may occur throughout the cerebral vasculature, they tend to occur at branch points in the circle of Willis and most commonly involve the middle cerebral vasculature. Identifying the aneurysm upon initial removal and inspection of the brain may be difficult because of large amounts of formed blood clots (Figure 11.7). When an aneurysm or vascular malformation is suspected, the subarachnoid membrane is removed,

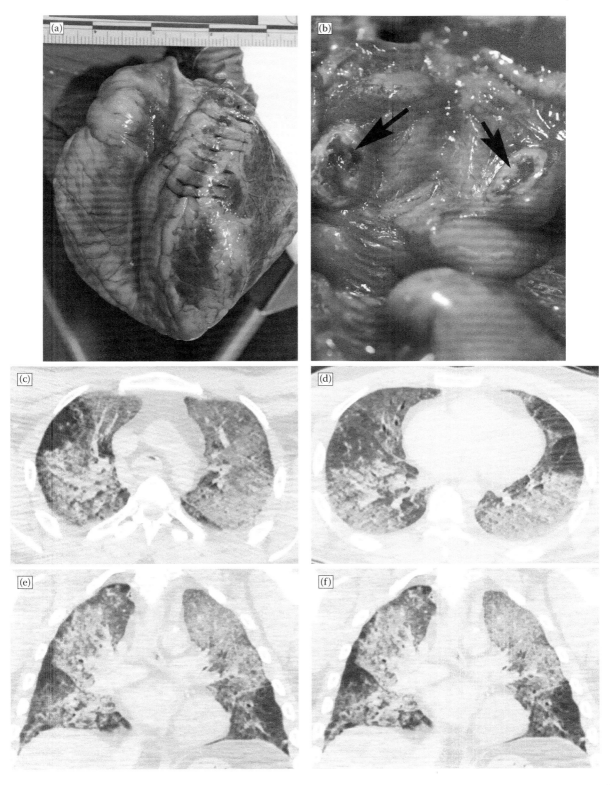

FIGURE 11.1   Acute myocardial infarction in a 53-year-old man. (a, b) Autopsy photograph of the heart shows sectioning of the left anterior descending artery to evaluate for atherosclerotic disease in (a). The left anterior descending coronary artery is opened in (b), showing occlusive thrombosis (arrows). Ninety percent narrowing was also present in the circumflex and right coronary arteries. (c, d, e, f) Axial and coronal MDCT image of the chest in lung window settings show diffuse, severe, pulmonary edema with septal lines, atherosclerotic calcification in the descending aorta, and no evidence of coronary artery calcification.

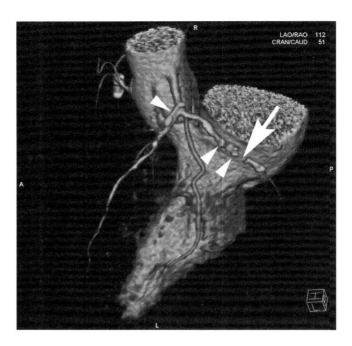

FIGURE 11.2   Volume-rendered MDCT angiography in a victim of acute myocardial infarction shows occlusion of the left circumflex artery (arrow) with overlying calcified plaque (arrowheads). (Courtesy of Dr. Steffen Ross, Institute of Forensic Medicine, Bern, Switzerland.)

the surface of the brain is washed, and the brain is fixed in formalin before dissection of the cerebral vasculature (Black and Graham 2002).

Cerebral infarction is an uncommon cause of sudden death. More commonly, death from cerebral infarction occurs within the hospital several days following infarction. Intracerebral hemorrhage accompanies up to 30% of infarctions. The mortality rate from cerebral infarction increases when hemorrhage is present, particularly if there is a large amount of hemorrhage. The hemorrhage may extend into the ventricles, subdural, or subarachnoid spaces.

Respiratory Disease

Although respiratory diseases account for approximately 10% of natural deaths, they are a very unusual cause of sudden death (Di Maio and Di Maio 2001). Pulmonary thromboemboli; acute airway obstruction from foreign bodies, epiglottitis, and asthma or other cases of bronchospasm; and massive hemoptysis from necrotizing pneumonias or tuberculosis can result in sudden death.

Pulmonary thromboembolism is a very common disease that is a major cause of morbidity and mortality in hospitalized patients. It also occurs in healthy individuals who are at risk for developing deep venous thrombosis

from immobilization, trauma, and hypercoaguable states. Death generally occurs when there is massive thromboembolism that occludes the main pulmonary arteries or central branches (Figure 11.8). Cardiovascular collapse and death in pulmonary embolism generally occur when more than 50% of the vasculature is occluded. At autopsy, emboli that occlude both pulmonary arteries are more frequently found than unilateral occlusion (Morgenthaler and Ryu 1995). If the embolus lodges at the bifurcation of the main pulmonary artery such that it occludes both right and left branches, it is referred to as a *saddle embolus*. Emboli in the large-caliber pulmonary vessels may have visible *lines of Zahn* that are created by laminations from layers of platelets mixed with fibrin alternating with layers containing more red blood cells (Kumar and Robbins 2007). The lung parenchyma may show minimal changes or may have significant congestion and hemorrhage. Pulmonary infarction may occur, especially in the setting of small emboli and if there is underlying compromise of the bronchial artery supply. Infarcts are classically hemorrhagic, raised, reddish blue areas that extend to the periphery of the lung and have an apex pointing toward the hilum. Fibrinous exudates may be found on the pleural surface. Acute dilatation of the right cardiac chambers (acute cor pulmonale) may also be found. In addition to

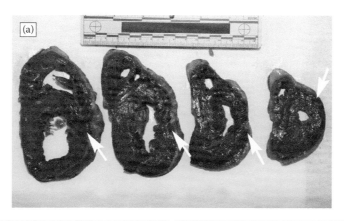

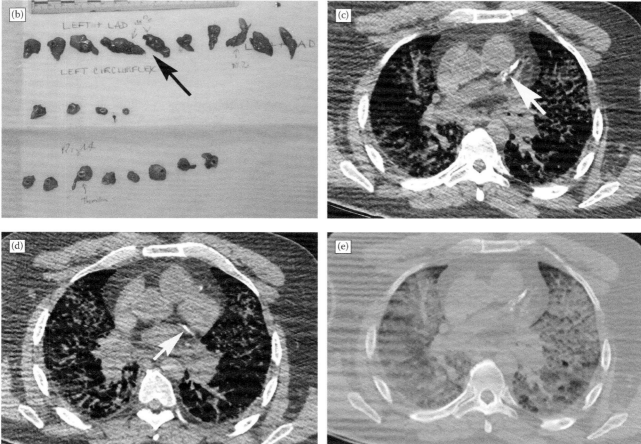

FIGURE 11.3   Acute myocardial infarction from atherosclerotic cardiovascular disease in a 63-year-old man. (a, b) Autopsy photographs of the sectioned heart and coronary arteries show myocardial necrosis (arrows) in the anterior left ventricle consistent with acute infarction from occlusion of the left anterior descending artery. The occluded left anterior descending artery (arrow) is sectioned in (b). (c, d, e) Axial MDCT image of the chest shows extensive atherosclerotic calcification in the left anterior descending coronary artery (arrow in c) and in the left circumflex artery (arrow in d). The bilateral pulmonary opacities represent alveolar pulmonary edema, which is best shown in lung window settings in (e).

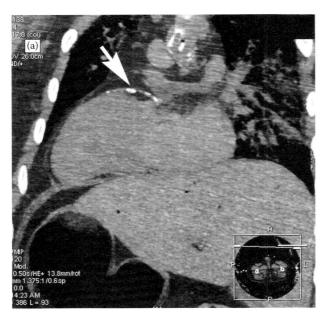
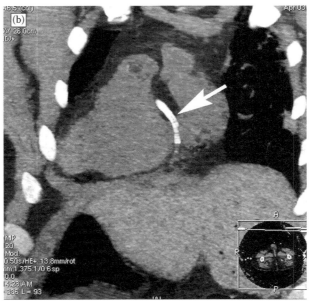

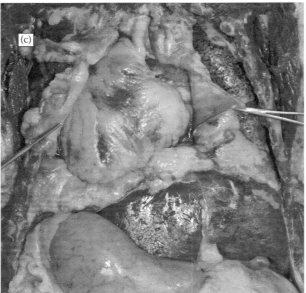

FIGURE 11.4 Acute myocardial infarction in a 55-year-old man with situs inversus totalis who had a prior history of coronary artery disease. (a, b) Coronal MDCT images of the chest and upper abdomen show situs inversus totalis. There are atherosclerotic calcifications in the "anatomic" left coronary artery (arrow in a) and two overlapping coronary artery stents (arrow in b) in the "anatomic" right coronary artery. (c) Autopsy photograph of the opened chest cavity shows situs inversus.

thromboemboli, other forms of pulmonary emboli may also be lethal. These include traumatic or iatrogenic air emboli, bone marrow emboli in severe skeletal trauma and sickle cell anemia patients, fat emboli from trauma and surgery, amniotic fluid emboli, and foreign bodies.

Deaths from pneumonia are more commonly seen in the hospital setting rather than in the medical examiner's office. However, on occasion, pneumonia is a cause of

unexplained death, particularly if the causative organism is highly virulent or if the decedent did not have access to medical care. Severe alveolar consolidation in pneumonia causes the lungs to be heavy, boggy, red, firm, and airless on gross examination. The term *hepatization* is often used to characterize the gross appearance of the lungs in severe pneumonia, because they have a liver-like consistency. Gray hepatization refers to the grayish brown appearance of

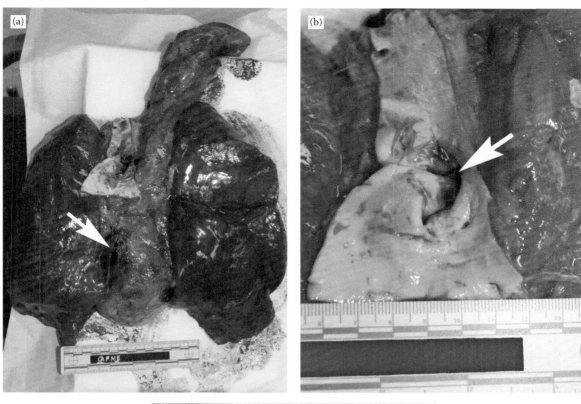

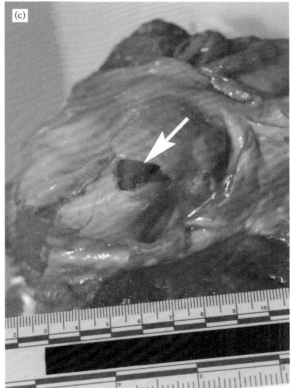

FIGURE 11.5   Acute aortic dissection in a 52-year-old man who was found unresponsive. (a, b, c) Autopsy photographs show a hemopericardium (arrow in a) and the opened aorta, which has a 5 cm V-shaped defect (arrow in b and c) in the aortic intima which leads to a false lumen.

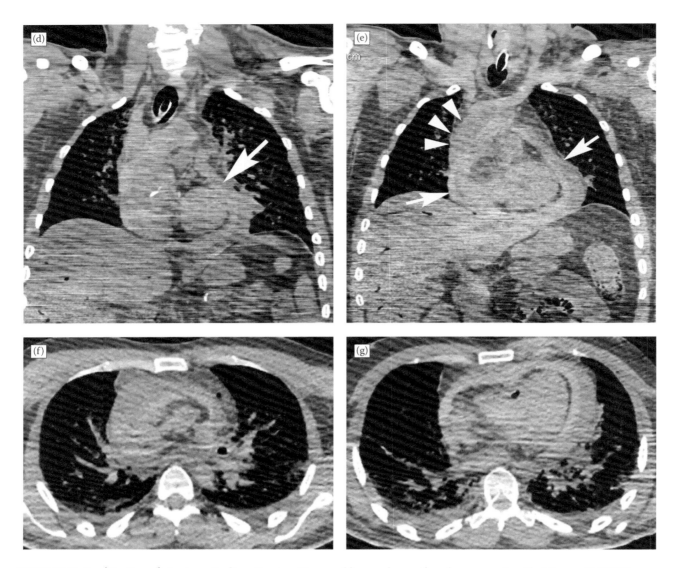

FIGURE 11.5 (*Continued*) Acute aortic dissection in a 52-year-old man who was found unresponsive. (d, e) Coronal MDCT images of the chest in soft tissue window show a large hemopericardium (arrows) and focal enlargement of the ascending aorta (arrowheads in e) with intramural high attenuation. An endotracheal tube is in the trachea. (f, g) Axial MDCT images show hemopericardium and bilateral alveolar densities from pulmonary edema.

lungs that have a fibronosuppurative exudate (Figure 11.9) (Kumar and Robbins 2007). Microbiology and viral cultures are necessary to establish the causative organisms.

Intra-Abdominal Hemorrhage

Unexpected death from intra-abdominal disease is uncommon. Massive gastrointestinal, intraperitoneal, and retroperitoneal hemorrhage may result in death. Exanguinating gastrointestinal hemorrhage from peptic ulceration, neoplastic disease, or variceal bleeding is unusual. Intra-abdominal hemorrhage may be caused by rupture of an aneurysm; splenic or hepatic rupture in the setting of an underlying infection, tumor, or as a delayed manifestation of blunt trauma; and rupture of a tubal pregnancy in women of childbearing age. The finding of large amounts of blood in the peritoneal cavity (Figure 11.10) or retroperitoneum at autopsy should lead to careful evaluation of the abdominal vasculature and sectioning of all visceral organs.

RADIOLOGIC PRINCIPLES
Goals of Imaging

The primary goal of postmortem imaging in a suspected natural death is to evaluate for occult trauma and to

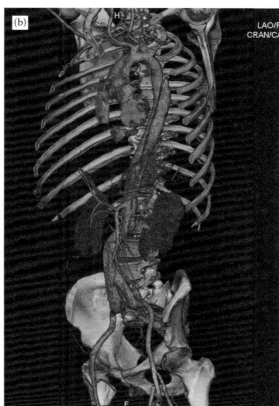

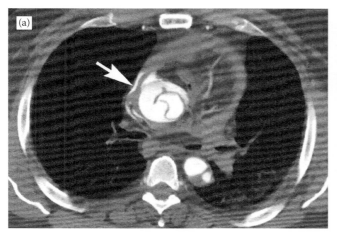

FIGURE 11.6 Postmortem MDCT angiography of acute aortic dissection. (a) Axial MDCT shows contrast in the true and false lumen of a DeBakey Type I aortic dissection of the ascending and descending aorta. There is a hemopericardium and extravasation of intravascular contrast material into the pericardium (arrow). (b) Volume-rendered MDCT shows the distal extent of the aortic dissection in the descending aorta. (Courtesy of Dr. Steffen Ross, Institute of Forensic Medicine, Bern, Switzerland.)

identify a potential cause of death. Postmortem multidetector computed tomography (MDCT) is the most appropriate initial cross-sectional imaging technique because it provides a rapid anatomic survey of the head and body. Other imaging techniques such as MDCT angiography and magnetic resonance imaging (MRI) may prove to be useful to clarify the imaging diagnosis or suspected diagnosis by autopsy if the autopsy findings are inconclusive.

Injuries from cardiopulmonary resuscitation should not be mistaken for traumatic injuries or disease. Resuscitation-associated rib and sternal fractures are the most commonly identified on the anterior portion of the ribs and typically do not have associated hemorrhage. They are typically anterior buckle fractures involving the inner cortex of the rib. Sternal fractures from resuscitation are most common in the lower portion of the sternum (Lederer et al. 2004). Rib fractures in the cartilaginous portion of the rib or at the costochondral junction may be found at autopsy but are not evident on MDCT. Gastric distension, pneumothorax,

pneumoperitoneum, and pneumoretroperitoneum may be caused by resuscitation. After excluding occult trauma, the MDCT images should be meticulously evaluated for anatomic findings that may support a specific cause of death.

Imaging Findings

Cardiovascular Disease — Postmortem MDCT is an excellent imaging modality for the detection and quantification of coronary artery calcification, which indicates the presence of atherosclerotic vascular disease, and for the evaluation of secondary findings such as pulmonary edema. However, the degree of coronal artery luminal narrowing and presence of superimposed thrombosis cannot be assessed without intravascular contrast material. The absence of coronary artery calcification does not exclude the presence of significant plaque or thrombotic occlusion of the coronary arteries (Figures 11.1 and 11.2). Furthermore, the tissue alterations that occur within the myocardium from infarction

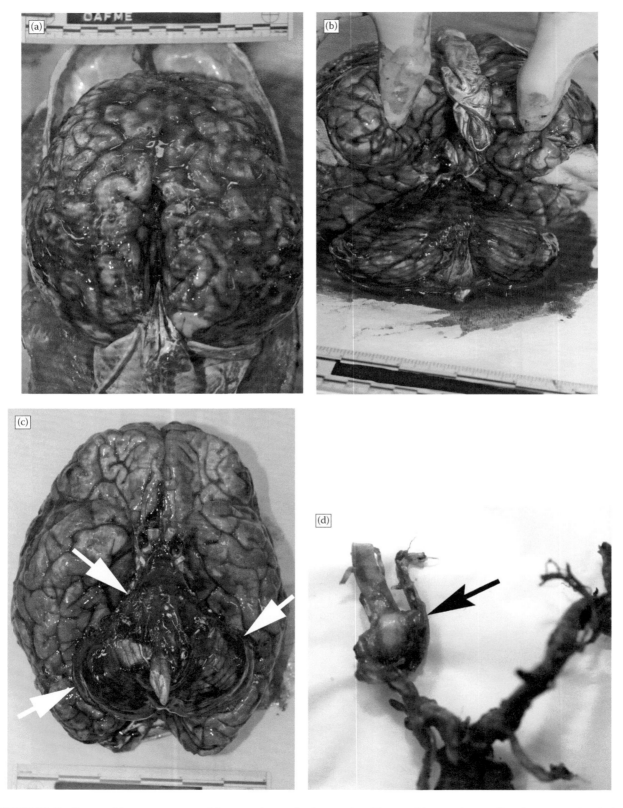

FIGURE 11.7   Ruptured berry aneurysm of the posterior cerebral artery in a 59-year-old man who was found unresponsive. (a, b, c) Autopsy photographs show diffuse subarachnoid hemorrhage and clotted hemorrhage surrounding the brain stem and cerebellum (arrow in c). (d) Photograph of the dissected circle of Willis shows an aneurysm of the posterior cerebral artery (arrow).

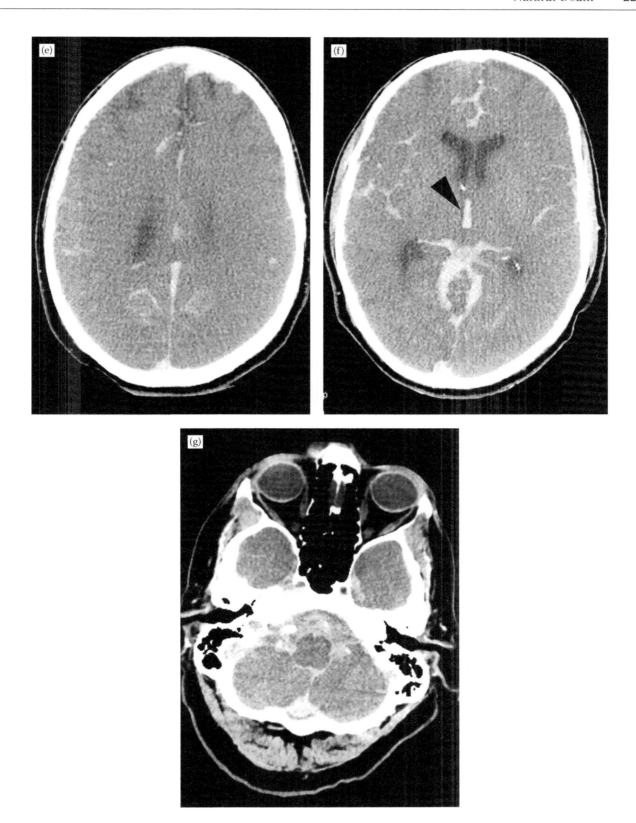

FIGURE 11.7 (*Continued*) Ruptured berry aneurysm of the posterior cerebral artery in a 59-year-old man who was found unresponsive. (e, f, g) Axial MDCT images of the brain show diffuse high-attenuation subarachnoid hemorrhage with extension of hemorrhage into the third ventricle (arrowhead in f). More focal hemorrhage is present in the quadrageminal plate cistern and premedullary cerebrospinal fluid space.

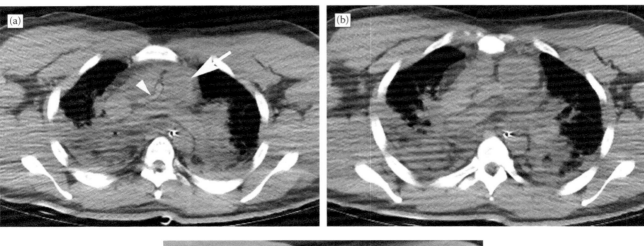

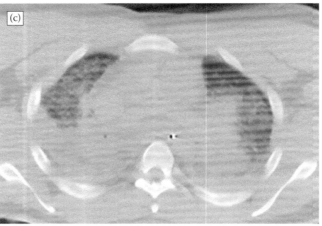

FIGURE 11.8 Bilateral pulmonary thromboembolism in a man who suddenly collapsed, was found unresponsive, and failed attempted resuscitation. (a, b) Axial MDCT of the chest in soft tissue window shows enlargement of the main pulmonary artery (arrow in a) adjacent to a collapsed ascending aorta (arrowhead in a). A nasogastric tube is in the esophagus. (c) Axial MDCT of the chest in lung window shows bilateral alveolar opacification, which was demonstrated to represent hemorrhage at autopsy. A nasogastric tube is in the esophagus.

are not detectable on MDCT. Consequently, noncontrast postmortem MDCT is not diagnostic in the determination of atherosclerotic coronary artery disease as a cause of death but provides supportive information. In our experience, the most common MDCT findings in death from myocardial infarction from atherosclerotic coronary artery disease are coronary artery calcification and pulmonary edema (Figure 11.3). Occasionally, other findings may be present that indicate the decedent has underlying cardiovascular disease. Pacemakers, surgical clips, coronary artery stents, and evidence of peripheral or cerebrovascular atherosclerotic disease may be present. The appearance of coronary artery stents on MDCT varies by manufacturer and stent composition.

Most stents have lines of parallel metallic attenuation on MDCT (Figure 11.4). Although routine postmortem MDCT may show fractures and malposition of coronary stents, the patency of stents cannot be assessed without intravascular contrast material.

Recent research has focused on MDCT angiography and MRI for postmortem cardiovascular evaluation. Postmortem angiographic techniques may be used for evaluation of the coronary arteries (Ross et al. 2008). Two- and three-dimensional MDCT images may be used to establish the degree of luminal narrowing or the presence of arterial occlusion when intravascular contrast material is injected into the arterial system during MDCT imaging

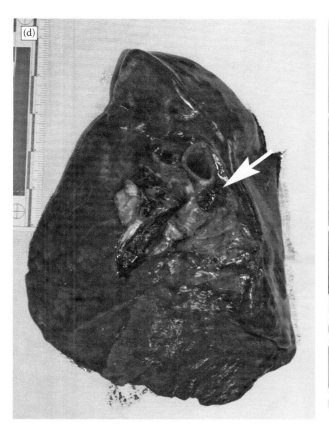
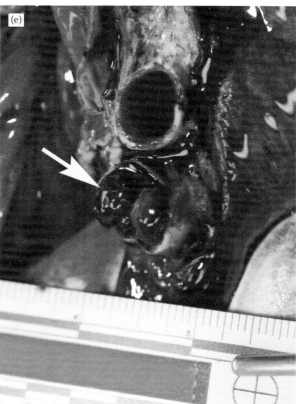

FIGURE 11.8 (*Continued*) Bilateral pulmonary thromboembolism in a man who suddenly collapsed, was found unresponsive, and failed attempted resuscitation. (d, e) Autopsy photographs of the left lung show congestion, hemorrhage, and an occluding thrombus in the main pulmonary artery (arrow in d and e).

(Figure 11.2). Arterial vessels with atherosclerotic disease have irregular luminal contours. Calcified and thrombotic plaque may have ulcerations, which appear as irregularities along the margins of the plaque.

On MRI, acute myocardial infarction shows central low signal intensity with surrounding hyperintensity on T2-weighted, short TI inversion recovery (STIR), and fluid-attenuated inversion recovery (FLAIR) sequences in the regions of the affected myocardium, whereas chronic myocardial ischemic change is typically low signal intensity on all pulse-weighted sequences (Jackowski et al. 2005, Jackowski et al. 2006). Hemorrhagic infarctions may also show increased signal intensity on T1-weighted images.

Deaths from aortic aneurysm rupture result from massive hemorrhage. Hemorrhage from intrathoracic aneurysm rupture may surround the aorta and extend into the mediastinum, pericardium, or pleural spaces. On MDCT, acute hemorrhage is higher in attenuation than intravascular blood. Aneurysmal dilatation of the aorta may or may not be apparent on MDCT, because if residual intravascular blood volume is low, the aorta may be collapsed on postmortem imaging. In some cases, it may be difficult to accurately identify the aorta because of collapse and adjacent hemorrhage.

Aortic dissection may occur with or without aneurysmal dilatation of the aorta. The postmortem MDCT findings in aortic dissection are deformity of the aortic contour, intramural hematoma, hemopericardium, and pulmonary edema (Figure 11.5). Hemopericardium, characterized by a hyperdense inner ring and hypodense outer ring or gravity-dependent hyperdensity from postmortem lividity, is the most common MDCT finding in deaths from aortic dissection (Figure 11.5g). This finding has been called the *hyperdense armored heart* by some authors (Shiotani et al. 2004). Angiography is required to confidently identify the

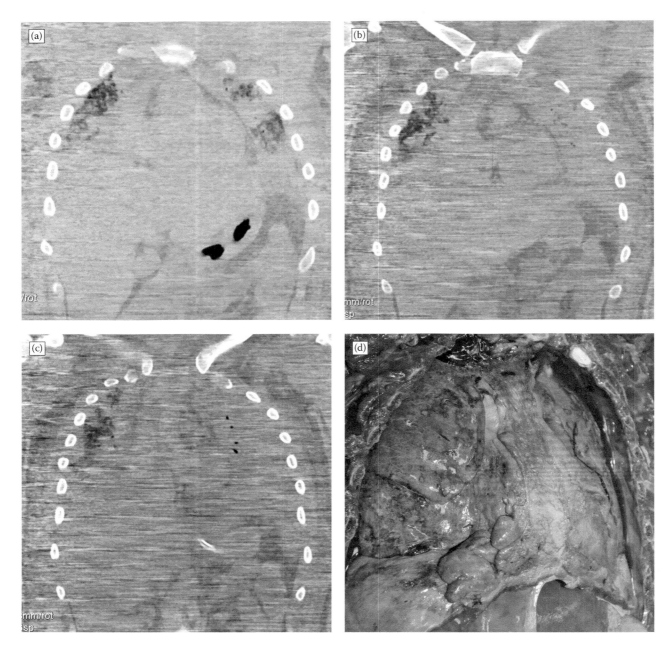

FIGURE 11.9   Fatal *Staphylococcus aureas* pneumonia. (a, b, c) Coronal MDCT images of the chest show near-complete alveolar opacification of both lungs. There is a small area of patchy aerated lung in the right upper lobe. Bilateral pleural effusions are shown in (b) and (c), but they are difficult to distinguish from the adjacent opacified lung parenchyma. Similarly, the hemidiaphragms are not well seen on the coronal image because of pulmonary opacification and effusions. A nasogastric tube is present in (c). (d) Autopsy photograph of the lungs *in situ* shows grayish brown, boggy, firm lungs with fibrinosuppurative exudates.

intimal flap and false lumen of the dissection when establishing the diagnosis by imaging alone (Figure 11.6).

Intracranial and Cerebrovascular Disease

To our knowledge, there have been no reports in the medical literature that describe the postmortem MDCT findings in sudden unexpected death in epilepsy. The role of postmortem imaging in these deaths is to exclude intracranial pathology that may be a seizure focus, such as occult trauma, hemorrhage, or tumor, and to exclude other causes of death. Tumors and cerebral infarctions are most often low attenuation on noncontrast

FIGURE 11.9 (*Continued*) Fatal *Staphylococcus aureas* pneumonia. (e) Photograph of the cut surface of the removed left lung shows dense, red, firm congestion.

postmortem MDCT. They may exhibit mass effect from vasogenic edema, and evidence of cerebral herniation may be present. It is uncommon for tumors to cause sudden death—the estimated incidence is between 0.16% and 3.2% (Buttner et al. 1999).

Acute intracranial hemorrhage is high attenuation (80 to 90 Hounsfield units) on MDCT because the protein in blood has a high-attenuation coefficient. During life, hemorrhage reaches its maximum attenuation in the first 2 to 4 hours at which time clot formation and retraction occur. If a decedent survives beyond the acute phase, hemorrhage becomes progressively lower in attenuation. Intraparenchymal hemorrhage will be surrounded by vasogenic cerebral edema, which reaches its maximum at 4 to 5 days. Over time, the margins of intraparenchymal hemorrhage become less distinct.

On MDCT, subarachnoid hemorrhage is high attenuation in the subarachnoid space. Diffuse subarachnoid hemorrhage is characterized by high attenuation throughout the subarachnoid spaces that interdigitates between the cerebral gyri and settles in the basilar cisterns (Figure 11.7). In the setting of sudden death, cerebral aneurysms are the most common cause of diffuse subarachnoid hemorrhage. Other etiologies include arteriovenous malformations, coagulopathies, hemorrhagic infarction, and hemorrhagic tumors. The predominant location of subarachnoid hemorrhage on MDCT may be a clue to the location of the aneurysm. For example, anterior communicating artery aneurysms may rupture into the cistern of the lamina terminalis and interhemispheric fissure; posterior communicating artery aneurysms rupture into the supracellar cistern, and blood may enter the ipsilateral temporal horn of the lateral ventricle; basilar tip aneurysms may rupture into the suprasellar and quadrageminal plate cisterns, and blood may enter third ventricle; and middle cerebral artery aneurysms may rupture into the ipsilateral sylvian fissure. The aneurysm cannot be identified on routine postmortem MDCT. Postmortem angiography has the potential of demonstrating the aneurysm and site of rupture.

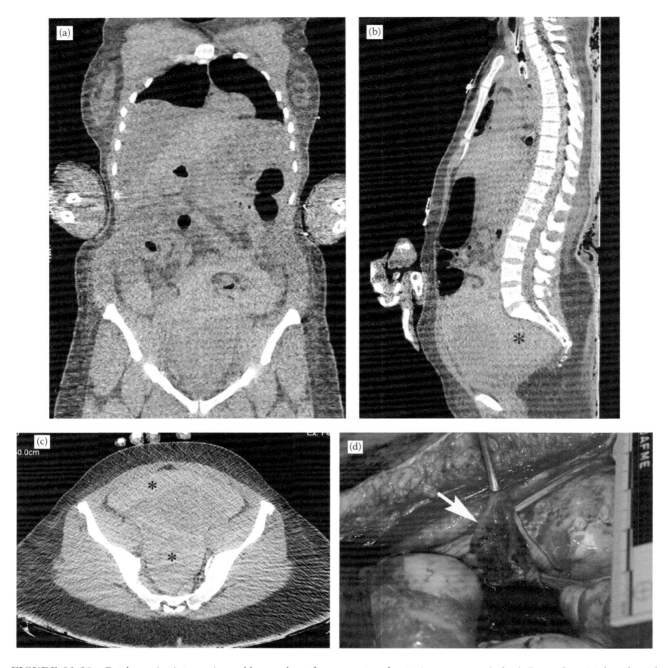

FIGURE 11.10   Fatal massive intraperitoneal hemorrhage from a ruptured ectopic pregnancy. (a, b, c) Coronal, sagittal, and axial MDCT show high-attenuation intraperitoneal hemorrhage with large amounts of clotted hemorrhage (asterisks in b and c) in the pelvis surrounding the bladder and uterus. (d) Autopsy photograph shows pelvic hemorrhage and a ruptured tubal pregnancy (arrow).

Subacute subarachnoid hemorrhage may not be detectable because its attenuation may begin to approximate that of cerebrospinal fluid. Subacute intraparenchymal hemorrhage may be iso- or hypoattenuating with the adjacent brain parenchyma, and the surrounding edema is hypoattenuating. At this stage on routine noncontrast postmortem MDCT, it may be impossible to distinguish between subacute intraparenchymal hemorrhage with edema from tumor or infarction. Imaging techniques such as MRI are more suited to distinguish between these entities.

Respiratory Disease

Postmortem MDCT has a small role in establishing the cause of death when it is respiratory in origin. Alveolar consolidation will be present in fatal pulmonary emboli and severe bronchopneumonia (Figures 11.8 and 11.9). Routine noncontrast postmortem MDCT is limited in the evaluation of pulmonary emboli, because intravascular contrast is necessary to identify pulmonary emboli. The main, right, and left pulmonary arteries may appear enlarged. There may be alveolar opacities from hemorrhage and infarction and small pleural effusions (Figure 11.8). Severe bronchopneumonia that results in death may show complete or near-complete opacification of the lungs on radiography and MDCT with associated exudative pleural effusions or empyemas (Figure 11.9).

Intra-Abdominal Hemorrhage

Acute intra-abdominal hemorrhage is high attenuation on postmortem MDCT (Figure 11.10). When interpreting scans, visual comparison of the attenuation of intra-abdominal fluid to the urine in the bladder may be an indication of the composition of the fluid collection. Blood is hyperattenuating compared to urine in the bladder, which is fluid attenuation. Intraperitoneal blood and fluid collect in the dependent recesses of the peritoneum: pouch of Douglas in women, retrovesical space in men, lateral paravesical recesses, paracolic gutters, and Morrison's pouch. The source of hemorrhage may not be detectable on unenhanced postmortem MDCT. In addition, the abdominal vasculature may be collapsed from severe hypovolemia. Consequently, detection of vascular abnormalities to include aneurysms is limited.

CONCLUSIONS

Postmortem MDCT in suspected natural deaths provides the medical examiner with information that can guide the autopsy and help evaluate for unsuspected injury or disease. Because routine postmortem MDCT is performed without intravenous contrast material, it provides a very limited assessment of cardiovascular disorders. However, the presence of focal or diffuse hemorrhage suggests the possibility of an underlying vascular etiology and may direct the autopsy if limited dissection is desired. Early research shows that postmortem MDCT angiography and MRI may improve the ability of postmortem imaging to noninvasively evaluate a suspected cardiovascular death.

REFERENCES

Black, M., and Graham, D. I. 2002. Sudden unexplained death in adults caused by intracranial pathology. *J Clin Pathol* 55: 44–50.

Buttner, A., Gall, C., Mall, G., and Weis, S. 1999. Unexpected death in persons with symptomatic epilepsy due to glial brain tumors: a report of two cases and review of the literature. *Forensic Sci Int* 100: 127–136.

Centers for Disease Control and Prevention. 2008. National center for health statistics. http://www.cdc.gov/nchs/fastats/deaths.htm (accessed October 16, 2008).

Christiansen, L. R., and Collins, K. A. 2007. Natural death in the forensic setting: a study and approach to the autopsy. *Am J Forensic Med Pathol* 28: 20–23.

de la Grandmaison, G. L. 2006. Is there progress in the autopsy diagnosis of sudden unexpected death in adults? *Forensic Sci Int* 156: 138–144.

de la Grandmaison, G. L., and Durigon, M. 2002. Sudden adult death: a medico-legal series of 77 cases between 1995 and 2000. *Med Sci Law* 42: 225–232.

Di Maio, V. J. M., and Di Maio, D. J. 2001. *Forensic pathology,* Boca Raton, FL: CRC Press.

Gonsoulin, M., Barnard, J. J., and Prahlow, J. A. 2002. Death resulting from ruptured cerebral artery aneurysm: 219 cases. *Am J Forensic Med Pathol* 23: 5–14.

Hutchins, G. M. 2003. Special studies on the heart and lungs. In Collins, K. A., and Hutchins, G. M. (Eds.), *Autopsy performance and reporting,* 2nd ed. Northfield, IL: College of American Pathologists.

Jackowski, C., Christe, A., Sonnenschein, M., Aghayev, E., and Thali, M. J. 2006. Postmortem unenhanced magnetic resonance imaging of myocardial infarction in correlation to histological infarction age characterization. *Eur Heart J* 27: 2459–2467.

Jackowski, C., Schweitzer, W., Thali, M. et al. 2005. Virtopsy: postmortem imaging of the human heart in situ using MSCT and MRI. *Forensic Sci Int* 149: 11–23.

Kumar, V., and Robbins, S. L. 2007. *Robbins basic pathology,* Philadelphia, PA: Saunders/Elsevier.

Lederer, W., Mair, D., Rabl, W., and Baubin, M. 2004. Frequency of rib and sternum fractures associated with out-of-hospital cardiopulmonary resuscitation is underestimated by conventional chest X-ray. *Resuscitation* 60: 157–162.

Morgenthaler, T. I., and Ryu, J. H. 1995. Clinical characteristics of fatal pulmonary embolism in a referral hospital. *Mayo Clin Proc* 70: 417–424.

Ross, S., Spendlove, D., Bolliger, S., Oesterhelweg, L., and Thali, M. 2008. Postmortem minimal invasive CT-angiography: the next step toward a virtual autopsy. *Radiologic Society of North America 94th Scientific Assembly and Annual Meeting.* Chicago, IL.

Shields, L. B., Hunsaker, D. M., Hunsaker, J. C., 3rd, and Parker, J. C., Jr. 2002. Sudden unexpected death in epilepsy: neuropathologic findings. *Am J Forensic Med Pathol* 23: 307–314.

Shiotani, S., Watanabe, K., Kohno, M. et al. 2004. Postmortem computed tomographic (PMCT) findings of pericardial effusion due to acute aortic dissection. *Radiat Med* 22: 405–407.

Chapter 12
Other Causes of Death

ASPHYXIA

Forensic Principles

Death from asphyxia is caused by cerebral hypoxia or anoxia. Asphyxial deaths are usually grouped into three broad categories: suffocation, strangulation, and chemical asphyxia. The autopsy findings of oxygen deprivation are nonspecific and include visceral congestion, petechiae, cyanosis, and fluidity of blood (Di Maio and Di Maio 2001).

The goal of forensic autopsy in asphyxial deaths is to discover findings that help to establish a specific cause of oxygen deprivation and to establish the manner of death as natural, accidental, suicide, or homicide.

Autopsy Findings

Suffocation — Suffocation may occur as a result of an environmental lack of oxygen, smothering from mechanical obstruction of the nose and mouth, choking from blockage of internal airways, or mechanical or traumatic asphyxia from external pressure on the chest or abdomen that restricts the ability to inspire. An individual trapped in an abandoned refrigerator is a classic example of environmental suffocation. There are no specific autopsy findings in these cases. The cause of death is generally established by excluding other causes of death and considering the circumstances surrounding the finding of the body in a locked chamber. Similarly, there are usually no specific autopsy findings in a victim of smothering. Petechiae may be present, but these are often considered nonspecific.

Mechanical airway obstruction is diagnosed at autopsy by finding an obstructing object such as food or a foreign body or in the case of acute epiglottitis, an enlarged and edematous epiglottis and inflammation of adjacent tissues. Mechanical asphyxia may be traumatic or positional. In traumatic causes, an accident has usually occurred that traps a person under the weight of an object that restricts breathing. Positional asphyxia occurs when a person is trapped in a position that prevents normal respiration. The autopsy in these cases may show visceral congestion, cyanosis, or petechiae. Nonlethal traumatic injuries may also be present.

Strangulation

Strangulation may occur from hanging, ligature compression of the neck, or manual compression of the neck. Death is from cerebral hypoxia secondary to carotid artery occlusion. Occlusion of the carotid arteries alone will cause death, because the vertebral arteries cannot provide enough blood flow to the brain (Di Maio and Di Maio 2001). The victim's face will be congested and cyanotic with numerous petechiae (Figure 12.1) when there is preservation of vertebral flow and occlusion of the carotid arteries and jugular veins, because the vertebral arteries continued to provide arterial flow to the brain without significant venous drainage. If all the major vessels are occluded such that no significant blood flow reaches the head, the victim's face appears pale. In complete vascular occlusion, the victim's tongue may protrude from the mouth and petechiae are generally absent.

Most hangings are suicides in which the body is partially or completely suspended by a noose around the neck. The noose produces a V-shaped furrow on the neck that is usually above the larynx (Figures 12.1 and 12.2). The markings and depth of the furrow depend upon the noose material. The knot of the noose represents the suspension point and produces a distinctive mark on the skin. Other autopsy findings include fractures of the hyoid bone, fractures of the thyroid or laryngeal cartilage, hemorrhage in the strap muscles, carotid artery injuries, pooling of blood and Tardieu spots in the lower extremities and forearms, and very rarely, cervical spine fractures (Di Maio and Di Maio 2001, Suarez-Penaranda et al. 2008). Cervical spine fractures are characteristic of judicial hangings in which the body drops from a height and is completely suspended by the noose. Death

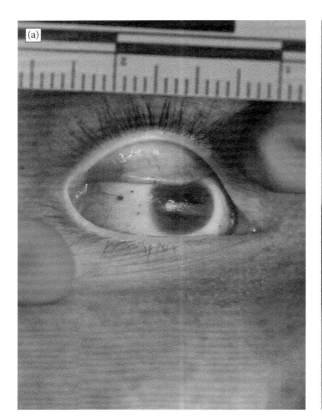

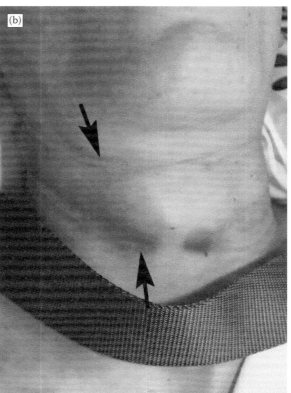

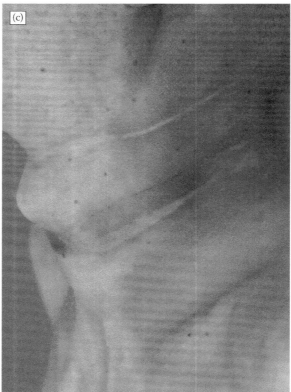

FIGURE 12.1   Suicide by hanging. (a) Autopsy photograph shows conjunctival petechiae. (b, c) Autopsy photographs show distinctive furrow marks (arrows) produced by both sides of a wide belt.

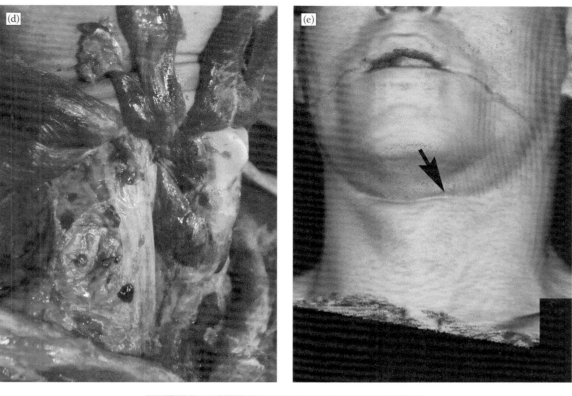

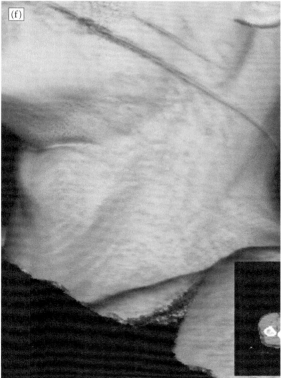

FIGURE 12.1 (*Continued*) Suicide by hanging. (d) Autopsy photograph of the dissected neck shows minimal hemorrhage in the neck muscles. (e, f) Surface-rendered three-dimensional MDCT images do not reliably reveal the furrow marks. The upper furrow is seen anteriorly beneath the chin (arrow) but is not visible laterally, and the lower furrow is not visible. On MDCT, skin folds beneath the chin can produce a crease that could be mistaken for a furrow mark.

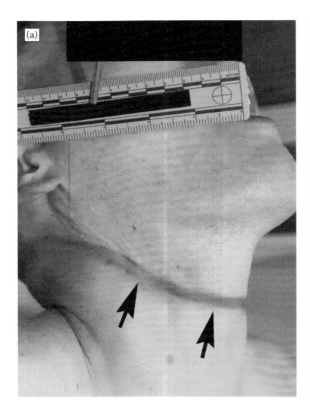

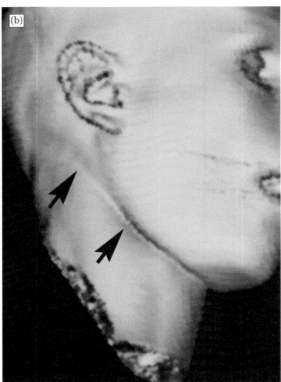

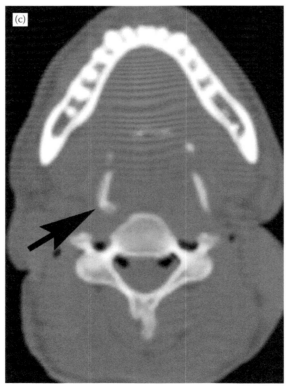

FIGURE 12.2 Suicide by hanging. (a) Autopsy photograph shows a deep V-shaped furrow (arrows). (b) Surface-rendered three-dimensional MDCT shows the furrow mark (arrows). It closely resembles a skin fold. Without comparison to the external examination or knowledge of the circumstances of death, the MDCT finding would be difficult to identify. (c) Axial MDCT shows an angulated fracture in the right greater cornu of the hyoid bone (arrow).

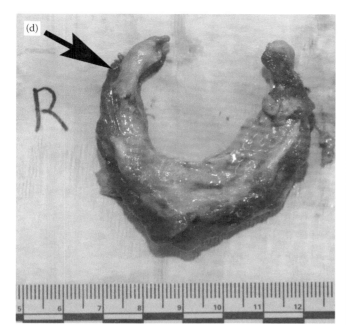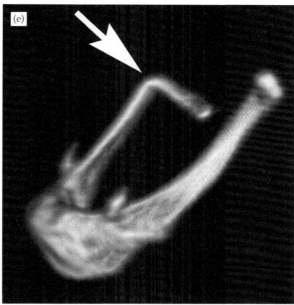

FIGURE 12.2 (*Continued*) Suicide by hanging. (d) Autopsy photograph of the removed hyoid bone shows the angulation of the right greater cornu at the site of the fracture (arrow). (e) Three-dimensional MDCT of the removed hyoid bone shows the angulated fracture (arrow).

in judicial hangings is caused by fracture dislocation of the cervical spine and transection of the spinal cord (Wallace et al. 1994).

Deaths from ligature and manual strangulations are most often homicides. Occasionally, accidental ligature strangulation occurs from neckties, scarves, or clothing around the neck that becomes entangled and tightened around the neck. The appearance and depth of ligature marks depends upon the type of ligature and the force applied. The victim's head and face are generally congested. The ligature and furrow marks are initially yellow and turn brown as they dry. In general, the marks are horizontally oriented. Hemorrhage in the soft tissues and fractures of the hyoid and thyroid cartilages may occasionally be found at autopsy. Manual strangulation occurs when an individual exerts pressure on the neck with his or her hands or forearms. Facial congestion, petechiae, abrasions, hemorrhages into the neck muscles and soft tissues, and hyoid and thyroid cartilage fractures may be found at autopsy.

Chemical Asphyxia
Chemical asphyxia occurs from inhalation of toxic gases that prevent cellular utilization of oxygen. Carbon monoxide, hydrogen cyanide, and hydrogen sulfide are the most commonly encountered chemical asphyxiants. Toxicology establishes the cause of death.

RADIOLOGIC PRINCIPLES

In the majority of asphyxial deaths, there are few radiologic findings. The role of imaging is to evaluate the airways, neck, and cervical spine for findings that help support the autopsy. Radiography and multidetector computed tomography (MDCT) are most useful in strangulation from hanging to identify laryngeal, hyoid, and cervical spine fractures. Following evaluation of the site of injury, imaging studies should be meticulously evaluated for other injuries. Recent published literature suggests that magnetic resonance imaging (MRI) may be helpful to show hemorrhage and edema in the soft tissues and cartilaginous structures of the neck in the forensic assessment of death by strangulation and in the clinical forensic assessment of victims that survive strangulation (Yen et al. 2005, Yen et al. 2007).

Imaging Findings
Suffocation and Chemical Asphyxia
Radiography and MDCT are used as adjuncts to autopsy in asphyxial deaths to aid in the identification of the source

of airway obstruction in suffocation from mechanical airway obstruction; identify associated traumatic injuries in mechanical or positional suffocation; and screen the body for unsuspected injury. There are no specific imaging features to aid in the diagnosis of smothering from mechanical obstruction of the nose and mouth and chemical asphyxia.

Strangulation

Hanging characteristically produces a V-shaped furrow around the neck because the body is suspended. The furrow coloration and associated abrasions or skin markings from the knot of the noose are findings that cannot be shown on MDCT. Furrow marks may be present on surface-rendered three-dimensional MDCT if they deform the skin surface such that they create an indentation in the skin (Figure 12.2). If there is no depression of the skin, the furrow mark will not be appreciated on MDCT (Figure 12.1). The furrow marks of hanging may be difficult to assess on MDCT because they closely resemble a skin fold beneath the chin. In all cases, the circumstances of death should be known prior to MDCT interpretation, and the findings should be correlated with physical examination. Similarly, the suspension point of the noose is difficult to identify on MDCT. It is usually a knot in the noose that creates a distinctive mark on the skin that is easily identified on external examination because its pattern is different than the remainder of the noose.

The value of MDCT in hanging deaths is to identify fractures of the hyoid, thyroid or laryngeal cartilage, and cervical spine. Hyoid fractures may be subtle and difficult to appreciate at autopsy when the entire hyoid is not removed. On MDCT, hyoid fractures appear as a discrete fracture line or as a focal deformity in the contour of the bone (Figure 12.2). Thyroid and laryngeal cartilage fractures are usually very difficult to identify on MDCT because the cartilage is usually not calcified. Cartilaginous calcification may be present in older individuals. Close observation of the pattern of calcification may reveal a disruption of the calcification, indicating a fracture.

Hangman's fracture of the C2 vertebral body is the classic fracture of judicial hangings. It is a fracture of both pedicles of C2 with anterior subluxation of C2. The posterior elements of C2 remain articulated with C3, and the anterior elements maintain their articulation with C1. Variations in C2 fractures may occur as well as fractures of the transverse processes, other upper cervical vertebrae, occipital bones, hyoid, and styloid processes. MDCT may also show disruption of the anterior and posterior longitudinal ligament with fracture dislocations, spinal cord transection, and in some cases, complete dissociation of the cervical spine (see Chapter 10, Figure 10.2). Spinal cord and ligament injury is best imaged with MRI. Findings such as disruption of the anterior and posterior longitudinal ligaments with associated intervertebral disk herniations, spinal cord transection, and edema and hemorrhage in the soft tissues of the neck have been reported on postmortem MRI of judicial hangings (Wallace et al. 1994).

ELECTROCUTION

The degree of injury from electrocution depends on the type of current, voltage, amperage, contact time, path of the current through the body, resistance of the body, and the presence or absence of moisture. Death may occur from high- or low-voltage circuits. High-voltage circuits, usually defined as those carrying more than 1000 volts, cause death from respiratory arrest or electrothermal injury (Di Maio and Di Maio 2001). Distribution and transmission power lines and lightning are sources of high voltage. Victims of high-voltage electrocutions almost always have electrical burns, which may be large confluent areas of full-thickness burns with charring. Multiple small burns may occur if the current arcs over the body. Death from low-voltage electrocution, less than 1000 volts, is from ventricular fibrillation. Burns may occur, but they tend to be small (Figure 12.3). If present, they are at the entry site of the electricity, usually the on the fingertips or hands, and at the exit site on the soles of the feet. When there are no burns in low-voltage electrocution deaths, there will be no specific autopsy findings. Careful investigation of the scene of death is necessary, along with testing and examination of electrical equipment from the scene to render a diagnosis based on the circumstances of death (Wright 1983).

To our knowledge, there is no published literature describing specific MDCT or MRI findings of

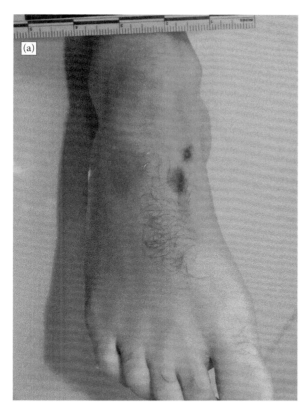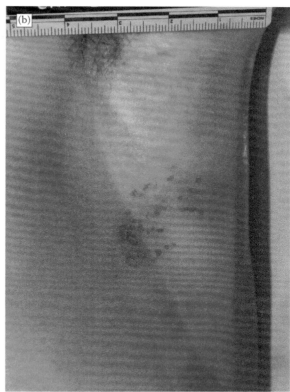

FIGURE 12.3 Electrocution burns. (a, b) Autopsy photographs show multiple small burns on the dorsum of the right foot and left lateral chest wall.

electrocution. In our experience, there are no correlative findings in electrocution when MDCT is compared to autopsy (Figure 12.3). MDCT is most useful in these cases to exclude occult trauma or disease that may be unsuspected. Although MRI may show soft tissue alterations from electrothermal injury, these wounds are optimally evaluated on direct examination at autopsy and microscopy.

CHILD ABUSE AND ELDER ABUSE

Child and elder abuse are important clinical and postmortem forensic topics that have significant societal impact. Both child and elder abuse encompass a wide spectrum of injuries that include physical trauma, sexual abuse, neglect, and psychological abuse. The radiologic and forensic findings in child abuse are well defined in the published medical literature since Caffey's 1946 publication on the radiologic manifestations of nonaccidental trauma in children (Caffey 1946, Dedouit et al. 2008, Kleinman 1998, Kleinman et al. 1989, Kleinman et al. 1995, Lonergan et al. 2003). More recently, elder maltreatment and abuse have

been recognized as a prevalent social problem (Collins 2006, Collins and Presnell 2006, Collins and Presnell 2007). The scope of these important causes of death is too large to cover in this text and has not been a part of our experience in postmortem MDCT. There is great potential for MDCT and MRI to make an impact in the clinical and postmortem assessment of abuse victims. Current published literature suggests that these imaging modalities are complementary to autopsy (Hart et al. 1996, McGraw et al. 2002, Thomsen et al. 1997).

CONCLUSIONS

MDCT is complementary in the forensic assessment of victims of strangulation. Important forensic findings such as subtle fractures of the hyoid and cervical spine are well assessed on computed tomography (CT). However, soft tissue findings such as furrow marks, abrasions, and bruising are not well evaluated. In electrocution, MDCT findings do not assist in determining electrocution as the cause of death but may be helpful in detecting occult or unsuspected injury.

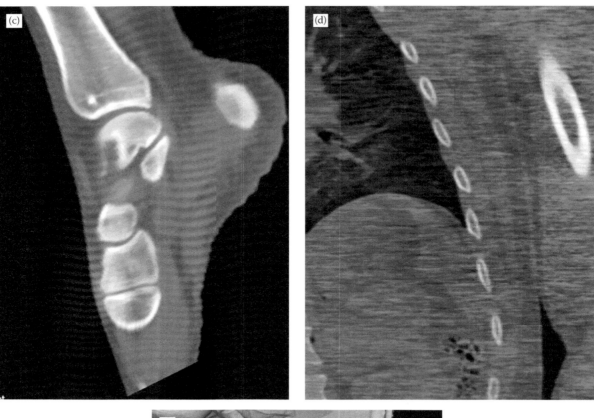

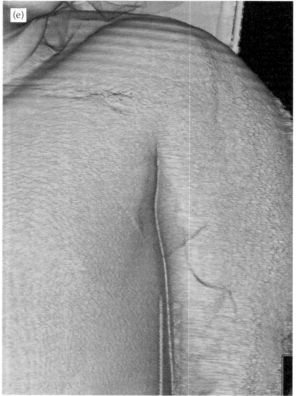

FIGURE 12.3   (*Continued*) Electrocution burns. (c, d, e) Sagittal MDCT of the right foot, coronal MDCT of the left chest, and surface-rendered three-dimensional MDCT of the left chest show no abnormalities.

REFERENCES

Caffey, J. 1946. Multiple fractures in the long bones of infants suffering from chronic subdural hematoma. *Am J Roentgenol Radium Ther* 56: 163–173.

Collins, K. A. 2006. Elder maltreatment: a review. *Arch Pathol Lab Med* 130: 1290–1296.

Collins, K. A., and Presnell, S. E. 2006. Elder homicide: a 20-year study. *Am J Forensic Med Pathol* 27: 183–187.

Collins, K. A., and Presnell, S. E. 2007. Elder neglect and the pathophysiology of aging. *Am J Forensic Med Pathol* 28: 157–162.

Dedouit, F., Guilbeau-Frugier, C., Capuani, C. et al. 2008. Child abuse: practical application of autopsy, radiological, and microscopic studies. *J Forensic Sci* 53: 1424–1429.

Di Maio, V. J. M., and Di Maio, D. J. 2001. *Forensic pathology,* Boca Raton, FL: CRC Press.

Hart, B. L., Dudley, M. H., and Zumwalt, R. E. 1996. Postmortem cranial MRI and autopsy correlation in suspected child abuse. *Am J Forensic Med Pathol* 17: 217–224.

Kleinman, P. K. 1998. *Diagnostic imaging of child abuse,* St. Louis: Mosby.

Kleinman, P. K., Blackbourne, B. D., Marks, S. C., Karellas, A., and Belanger, P. L. 1989. Radiologic contributions to the investigation and prosecution of cases of fatal infant abuse. *N Engl J Med* 320: 507–511.

Kleinman, P. K., Marks, S. C., Jr., Richmond, J. M., and Blackbourne, B. D. 1995. Inflicted skeletal injury: a postmortem radiologic-histopathologic study in 31 infants. *AJR Am J Roentgenol* 165: 647–650.

Lonergan, G. J., Baker, A. M., Morey, M. K., and Boos, S. C. 2003. From the archives of the AFIP. Child abuse: radiologic-pathologic correlation. *Radiographics* 23: 811–845.

McGraw, E. P., Pless, J. E., Pennington, D. J., and White, S. J. 2002. Postmortem radiography after unexpected death in neonates, infants, and children: should imaging be routine? *AJR Am J Roentgenol* 178: 1517–1521.

Suarez-Penaranda, J. M., Alvarez, T., Miguens, X. et al. 2008. Characterization of lesions in hanging deaths. *J Forensic Sci* 53: 720–723.

Thomsen, T. K., Elle, B., and Thomsen, J. L. 1997. Post-mortem radiological examination in infants: evidence of child abuse? *Forensic Sci Int* 90: 223–230.

Wallace, S. K., Cohen, W. A., Stern, E. J., and Reay, D. T. 1994. Judicial hanging: postmortem radiographic, CT, and MR imaging features with autopsy confirmation. *Radiology* 193: 263–267.

Wright, R. K. 1983. Death or injury caused by electrocution. *Clin Lab Med* 3: 343–353.

Yen, K., Thali, M. J., Aghayev, E. et al. 2005. Strangulation signs: initial correlation of MRI, MSCT, and forensic neck findings. *J Magn Reson Imaging* 22: 501–510.

Yen, K., Vock, P., Christe, A. et al. 2007. Clinical forensic radiology in strangulation victims: forensic expertise based on magnetic resonance imaging (MRI) findings. *Int J Legal Med* 121: 115–123.

Chapter 13

Beyond Standard Autopsy
More Roles for Postmortem Multidetector Computed Tomography

INTRODUCTION

Postmortem multidetector computed tomography (MDCT) affords an opportunity to address questions of medical and scientific interest apart from the primary role in supporting autopsy in determination of cause and manner of death. Radiology has traditionally played a role in identification of the dead by comparing antemortem and postmortem radiographs. The speed and ease of use of postmortem MDCT together with its three-dimensional capability makes it an even more valuable resource to answer specific questions of identification. Its exquisite spatial resolution makes it an ideal modality to apply to the study of human bones and decomposed remains. Anthropologists have already employed computed tomography (CT) in the study of ancient remains (Recheis et al. 1999). In our practice, MDCT has been used to evaluate specific questions about medical intervention and to assist forensic pathologists and anthropologists with the study of skeletal remains and exhumed bodies. In this chapter, we discuss the use of MDCT in the assessment of medical intervention, exhumation and second autopsy, and anthropology.

ASSESSMENT OF MEDICAL INTERVENTION

Victims of traumatic or natural deaths may be resuscitated or have an attempted resuscitation at the scene of the event, during transport, or upon reaching a treatment facility. Cardiopulmonary resuscitation requires the placement of various tubes, lines, catheters, and assistive medical devices (Figure 13.1). The documentation of intravascular lines, chest tubes, endotracheal or nasotracheal tubes, airway devices, tracheostomies, and nasogastric tubes, as well as less common devices such as intraosseous infusion catheters, is part of the autopsy external examination. MDCT adds to this assessment by providing multiplanar two- and three-dimensional images of the positions of these devices, because the internal positionings of most tubes and lines are easily established. The forensic pathologist can be alerted of malposition and may elect further assessment during dissection, if necessary. This is particularly helpful if first responders and other emergency personnel are using a new device or if there is a question regarding the position of a particular device. We emphasize that MDCT assesses device position at the time of the postmortem CT only because devices can shift position during transport and handling of the body. In addition, the clinical effectiveness of a particular device cannot be assessed because there are complex physiologic factors involved in injury and resuscitation. However, by providing feedback with regard to the anatomic positioning and physical characteristics, inferences can be made as to the effectiveness of a specific device or intervention (Harcke et al. 2007).

Intraosseous infusion catheters are well suited for MDCT assessment because the position of the tip of the intraosseous catheter cannot be assessed at routine autopsy. Intraosseous access is used in life-threatening emergencies when intravenous access using peripheral or central veins is unsuccessful (Luck et al. 2009). Cardiac arrest from hypovolemic shock, severe burn injury with shock, multiple traumas with shock, and status epilepticus are conditions that may require intraosseous access during resuscitation. These catheters are most often placed in the medullary space of the tibia (Figure 13.2) or sternum (Figure 13.3).

Placement and position of intraosseous infusion catheters is easily assessed with axial or three-dimensional MDCT. The catheter is correctly placed when its tip is in

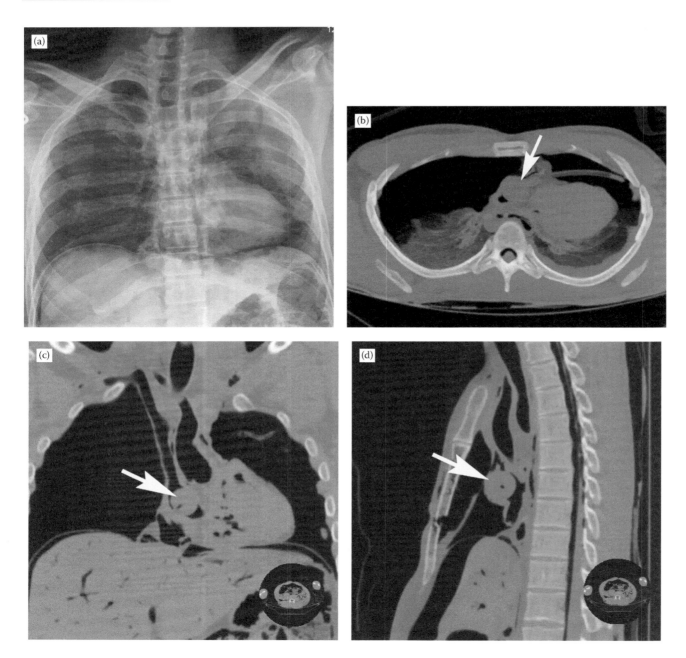

FIGURE 13.1 Emergency resuscitation surgical placement of a Foley catheter balloon in the right atrium. A Foley catheter can be used to control bleeding when the repair of a lacerated cardiac chamber is undertaken. (a) Postmortem radiograph shows bilateral pneumothorax, soft tissue air associated with a left thoracotomy, air in the right heart, and a pneumoperitoneum. (b, c, d) Axial, coronal, and sagittal air show the fluid-filled balloon of the catheter in the upper right atrium (arrows). Air is present in other spaces and structures and as described in (a).

the medullary cavity of the bone. Incorrect catheter placement may occur if the tip remains in the cortical bone or is located adjacent to the bone, missing the bone completely (Figures 13.4 and 13.5). Various tubes, lines, and medical devices can be assessed on postmortem MDCT in this manner. There is tremendous potential for further study of the position of medical devices and possibly the

effectiveness of medical intervention if anatomic data are combined with physiologic data.

EXHUMATION AND SECOND AUTOPSY

Second autopsy can be performed before burial or after exhumation. The success of an exhumation autopsy is

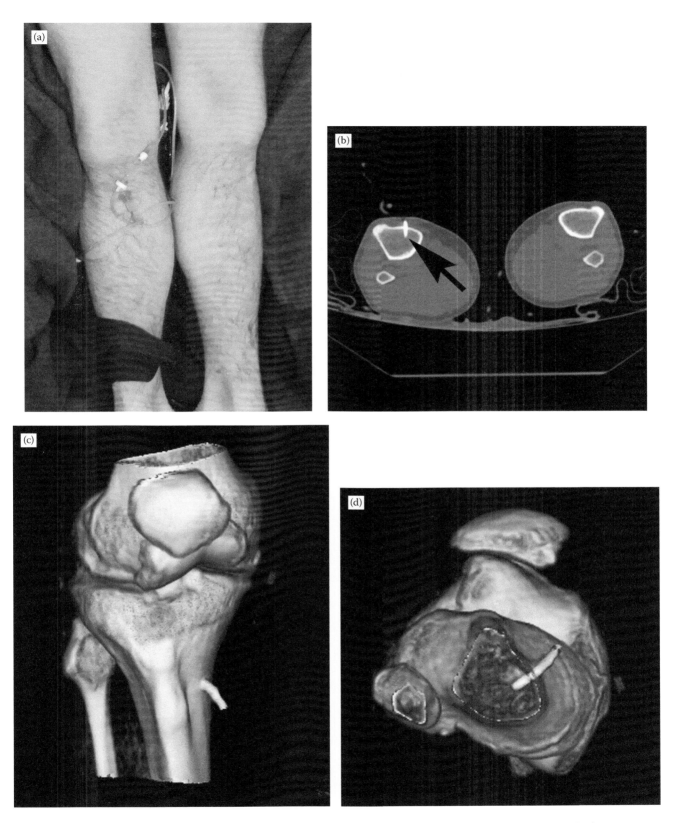

FIGURE 13.2 Correct position of a tibial intraosseous infusion catheter. (a) Autopsy photograph shows the tibial intraosseous catheter is placed in the proximal tibia, just medial and distal to the tibial tuberosity. (b) Axial MDCT shows the tip of the catheter (arrow) is located in the medullary space of the tibia. (c, d): Three-dimensional MDCT confirms the location in the medullary space.

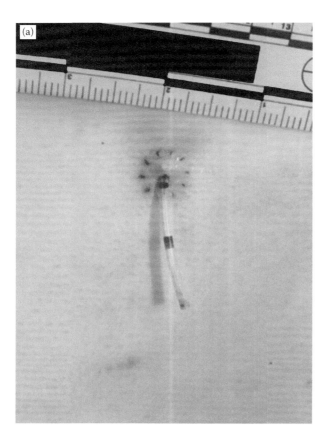

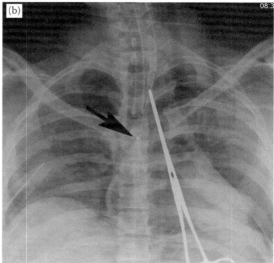

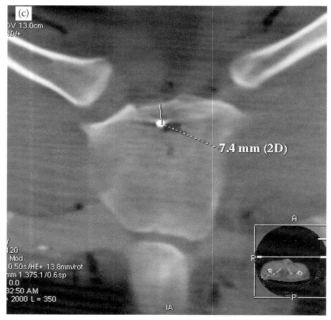

FIGURE 13.3 Sternal intraosseous intravenous infusion catheters. (a) Autopsy photograph shows an indwelling catheter exiting from the anterior chest wall. The circle of tiny punctuate lacerations is created by the insertion template device. (b) Postmortem chest radiograph shows a small metal density in the midline over T4 (arrow). This is the metal tip of the infusion catheter. Note the endotracheal tube, right chest tube, esophageal tube, and hemostat (outside the body). Air is present in the right heart. (c, d) Coronal and sagittal MDCT confirm the catheter tip is in the manubrium of the sternum. Distance from the superior margin of the manubrium is displayed.

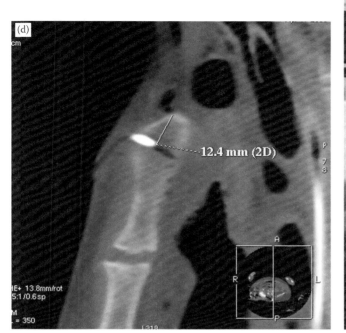

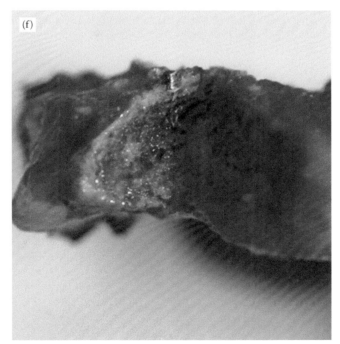

FIGURE 13.3 (*Continued*) Sternal intraosseous intravenous infusion catheters. (c, d) Coronal and sagittal MDCT confirm the catheter tip is in the manubrium of the sternum. Distance from the superior margin of the manubrium is displayed. (e, f) Autopsy photographs show placement of the metallic tip in a resected manubrium.

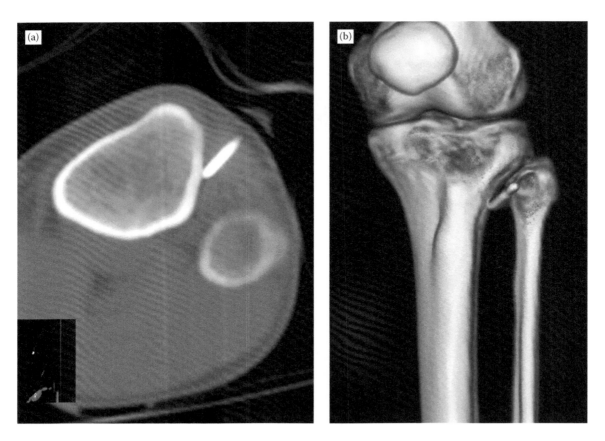

FIGURE 13.4   Incorrect position of a tibial intraosseous infusion catheter. (a) Axial and (b) three-dimensional MDCT show that the catheter did not enter the bone. The tip lies along the lateral aspect of the proximal tibia.

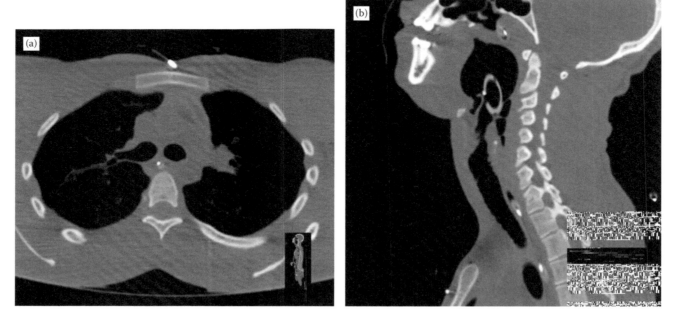

FIGURE 13.5   Incorrect position of a sternal intraosseous infusion catheter. (a) Axial and (b) sagittal MDCT show the metal tip of a sternal infusion catheter to be outside bone in the soft tissue overlying the manubrium. An airway device with an inflated balloon and a nasogastric tube is partially visible.

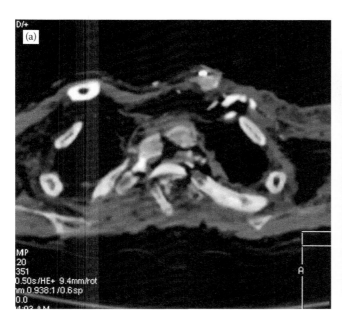
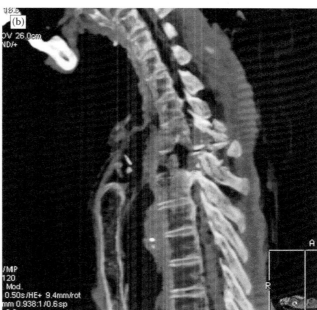

FIGURE 13.6 Forensic investigation of exhumed human remains revealed evidence of multiple gunshot wounds and a retained bullet fragment. (a, b) Axial and sagittal MDCT show fractures of left clavicle and T1 and T2 vertebral elements.

variable and related to the amount of decomposition. The success of a second autopsy is limited because the viscera have been previously dissected and examined. However, gross examination, microscopic tissue evaluation, and toxicology are possible many years after burial (Breitmeier et al. 2005, Grellner and Glenewinkel 1997). The amount of decay and decomposition in an exhumed body is most strongly associated with the length of time since death (Breitmeier et al. 2005). Other factors affecting the degree of decomposition and decay include environmental conditions such as the burial site, soil conditions, integrity of the coffin, and season of death. Bones, connective tissues, fat, and slowly decomposing organs such as the heart often retain evidence that may help ascertain the cause of death after exhumation (Figure 13.19). Mummification and adipocere formation (see Chapter 3, Figure 3.12) preserve anatomic structures and pathologic findings. Skin and subcutaneous findings such as bullet holes may be preserved for long periods of time in adipocerous bodies (Knight 1996).

MDCT facilitates the anatomic evaluation of exhumed bodies by providing a noninvasive anatomic assessment. If no autopsy was performed prior to burial, MDCT assists in localization of key anatomic organs and skeletal structures that may have shifted position during the decay

process. This can be particularly helpful in intermediate to advanced stages of decay. If the body has been previously autopsied, a plastic bag containing visceral organs is usually present in the body cavity. The bag can be radiographed and scanned separately from the body to prevent confusion when interpreting the images (Figure 13.7). The organ bag contains all organs and structures that were removed during the original autopsy dissection. Metallic fragments and bones may be located within the organs and tissues contained in the bag. They may have findings that are important to determining the cause of death. For example, the hyoid bone is usually removed during dissection of the neck, trachea, and larynx. Occult hyoid fractures may indicate that the victim sustained trauma to the neck (Figure 13.7). This additional information may be very helpful to the final determination of the cause of death.

ANTHROPOLOGY

Anthropologists study found human remains for age, gender, stature, and ethnicity by analyzing anthropomorphic features and osteometric criteria, which may help to establish identity. Unless purposefully buried deeply and intact, skeletal remains are commonly found dissociated. They may be scattered over a broad area from wildlife predation or purposely separated to cover a crime. When

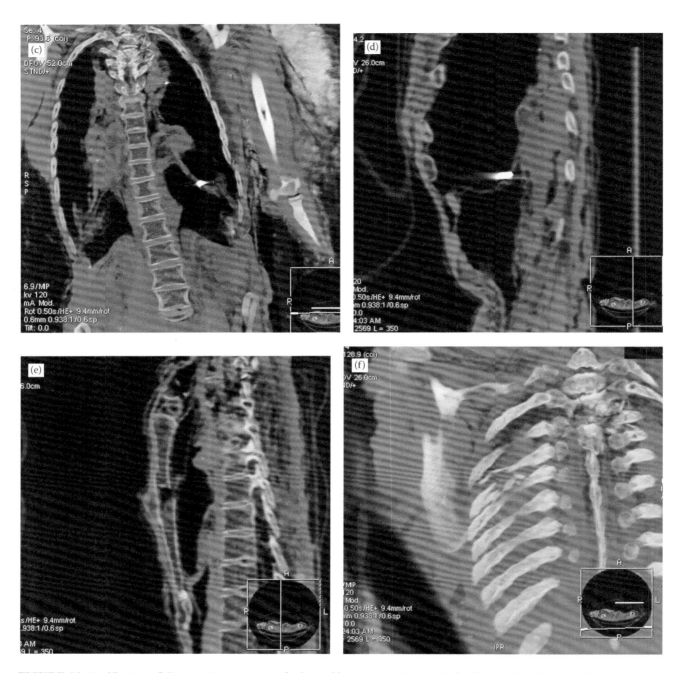

FIGURE 13.6 (*Continued*) Forensic investigation of exhumed human remains revealed evidence of multiple gunshot wounds and a retained bullet fragment. (c, d) Coronal and sagittal MDCT show a recoverable metal fragment in the lower left hemothorax. Tiny metal fragments are also noted elsewhere but are not likely to be recoverable. (e, f) Sagittal and coronal maximum intensity projection MDCT show a fracture defect in the sternum and comminuted right posterior rib fracture defect typical of a gunshot wound.

individual bones are found, they are analyzed for evidence of traumatic injury prior to death. Expert forensic anthropologists attempt to distinguish antemortem trauma from postmortem trauma and predation marks by characteristic features and locations of fractures or marks on the bone.

Reassembly of skeletal remains helps to further identify and classify findings that may help indicate the cause of death. Radiography and MDCT scanning of the reassembled remains is superior to imaging bones individually or collectively. Images of a partially or completely reassembled skeleton better depict injury patterns and show

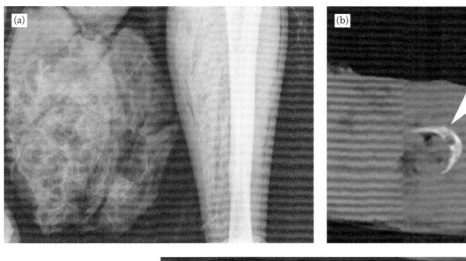

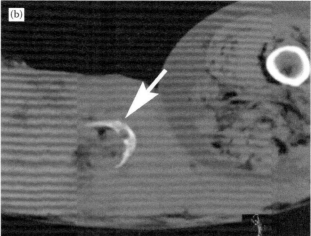

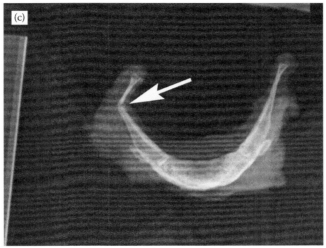

FIGURE 13.7 Second autopsy performed for medicolegal reasons. (a) Radiograph of the organ bag adjacent to the leg shows nonspecific soft tissue and gas. No metallic fragments or definite bones are identified. (b) Axial MDCT of the organ bag shows that the hyoid bone (arrow) is located within the contents of the bag. (c) Magnification radiograph of the recovered hyoid bone shows a fracture of the right cornu (arrow), which was not found on the first autopsy.

the relationship of wounds to one another (Figure 13.8). MDCT has also been used to aid in measurement of osteometric criteria and anthropomorphic analysis (Dedouit et al. 2007, Recheis et al. 1999, Verhoff et al. 2008) and to study mummies (Cesarani et al. 2003, Hoffman et al. 2002).

Human remains and dissociated body parts may be found during the investigation of catastrophic trauma or large-scale disaster events such as hurricanes, floods, mass transit accidents, aircraft accidents, and explosions. Recovery and identification of all human remains is one of the most important objectives of these investigations. In addition, forensic pathologists may be asked to determine if

cause of death was related to the event circumstances or was a homicide. MDCT has been proposed as a means of autopsy triage (Figure 13.9). Associating all remains from an individual is important not only for the investigation but also for family members. In most instances, DNA analysis and forensic dentistry are performed to confirm the identity of dissociated body parts. On occasion, body parts may be matched when specific bony landmarks or features are identified on MDCT (Figure 13.10). This technique requires careful observation of fracture fragment patterns to piece the fragments together. Once a possible match is identified on MDCT, the parts must be physically assessed to confirm the match. Although complex and time consuming, this manner of matching body parts

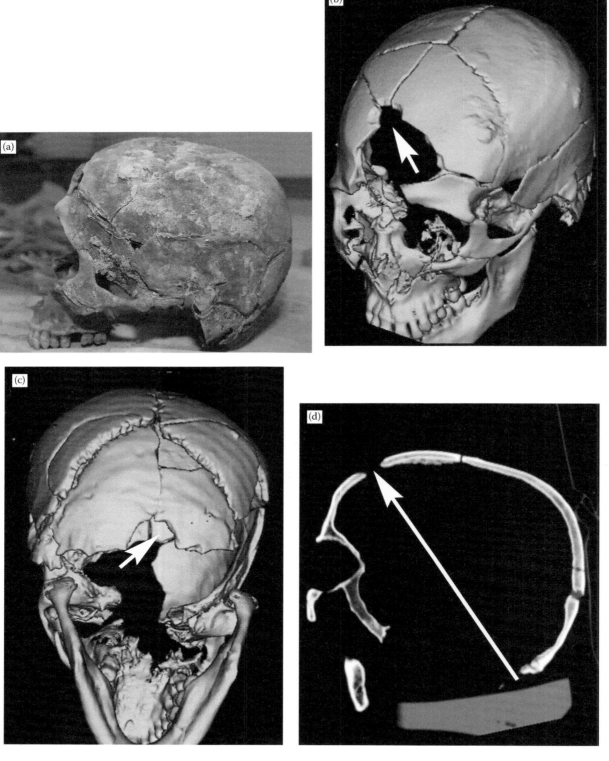

FIGURE 13.8   Forensic investigation of found skeletal remains shows findings of an execution type of gunshot wound to the head. (a) Photograph of the reassembled skull. (b) Three-dimensional MDCT of the front of the skull shows an exit defect with external beveling at the margin of a missing fragment (arrow). (c) Three-dimensional MDCT of the base of the skull shows an entrance defect at the margin of a missing fragment (arrow). (d) Sagittal oblique MDCT in the plane of the wound track (arrow) shows beveling of the entrance and exit wounds.

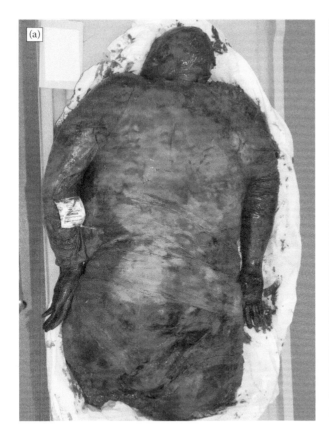
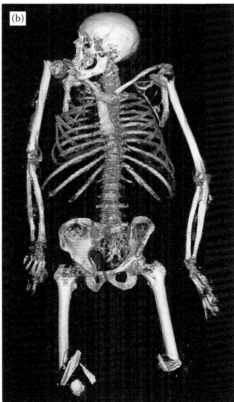
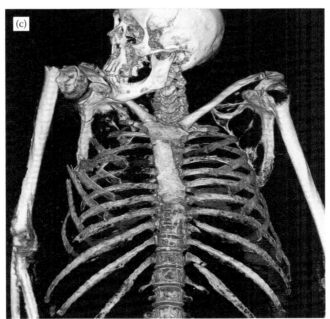

FIGURE 13.9   Decomposed remains recovered from a collapsed building at a disaster location. (a) Prone autopsy photograph shows decomposition changes in coloration and bloating. Note traumatic amputation of the lower extremities. (b, c) Three-dimensional MDCT reconstructions of the skeleton show fractures of the right humerus, sternum, multiple right and left ribs, and distal femurs. The reconstruction algorithm does not detect thin areas of flat bone such as the scapulas and iliac crests. As a result, these areas appear as holes in the bone.

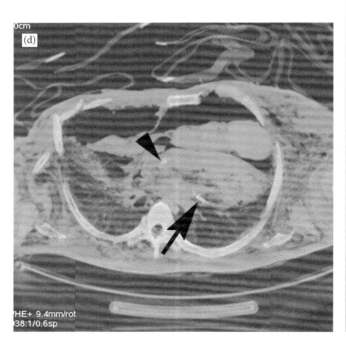

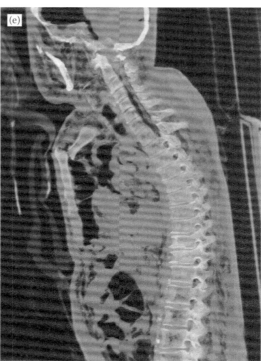

FIGURE 13.9 (*Continued*) Decomposed remains recovered from a collapsed building at a disaster location. (d, e) Axial and sagittal MDCT show the rib and sternal fractures consistent with blunt force injury. In the heart, calcification is present in the left coronary artery (arrow), and a stent is visible in the right coronary artery (arrowhead). The presence of the stent is notable because it could be used to help establish identification. Tissue and organ air distribution is typical of decomposition.

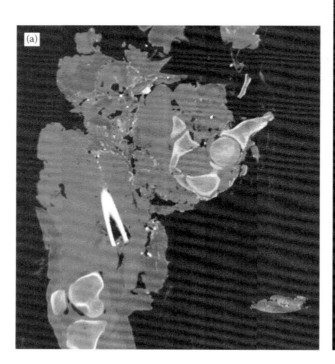

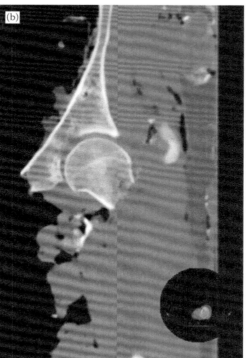

FIGURE 13.10   Forensic investigation of dissociated body parts using three-dimensional MDCT. (a, b) Coronal MDCT shows the dissociated parts each contain pelvic fragments; pubic symphysis (a) and acetabulum/hip joint.

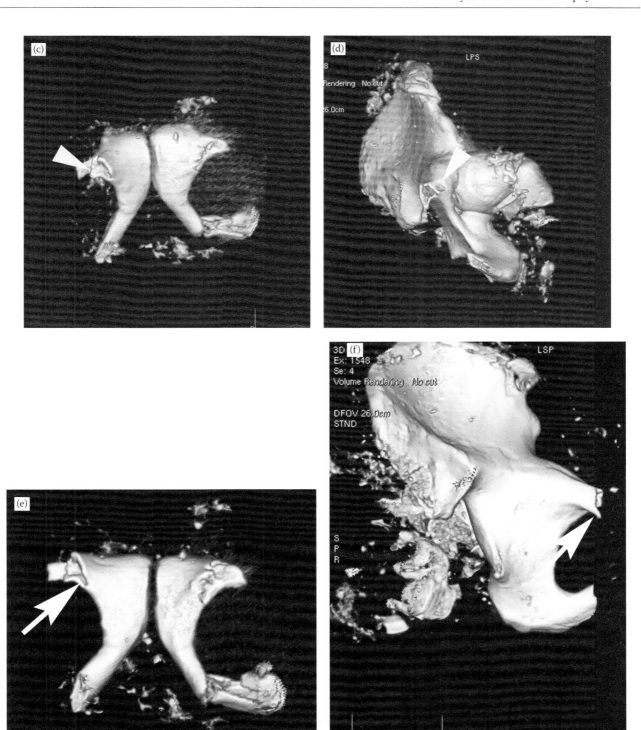

FIGURE 13.10 (*Continued*) Forensic investigation of dissociated body parts using three-dimensional MDCT. (c, d, e, f) Three-dimensional MDCT of pelvic fragments show each contain a matching bone feature (arrows) indicating that the two fragments belong together. Arrowheads in (c) and (d) and arrows in (e) and (f) point to complimentary bone features that indicate fragments belong together. MDCT reassociation obviated the need for DNA testing.

may be helpful in assisting forensic anthropologists or when DNA testing is limited or not available.

CONCLUSIONS

Throughout this textbook, we discussed and illustrated the use of postmortem MDCT in routine forensic autopsy. Postmortem MDCT may also be used to study specific anatomic areas, answer medical questions, evaluate medical intervention, complement nontraditional autopsy scenarios such as exhumation and second autopsy, and assist with forensic anthropology and disaster recovery. Postmortem MDCT and other advanced cross-sectional imaging modalities have enormous potential to assist forensic autopsy by adding objective anatomic data that can be reviewed before, during, and after autopsy.

REFERENCES

Breitmeier, D., Graefe-Kirci, U., Albrecht, K. et al. 2005. Evaluation of the correlation between time corpses spent in in-ground graves and findings at exhumation. *Forensic Sci Int* 154: 218–223.

Cesarani, F., Martina, M. C., Ferraris, A. et al. 2003. Whole-body three-dimensional multidetector CT of 13 Egyptian human mummies. *AJR Am J Roentgenol* 180: 597–606.

Dedouit, F., Telmon, N., Costagliola, R. et al. 2007. Virtual anthropology and forensic identification: report of one case. *Forensic Sci Int* 173: 182–187.

Grellner, W., and Glenewinkel, F. 1997. Exhumations: synopsis of morphological and toxicological findings in relation to the postmortem interval. Survey on a 20-year period and review of the literature. *Forensic Sci Int* 90: 139–159.

Harcke, H. T., Pearse, L. A., Levy, A. D., Getz, J. M., and Robinson, S. R. 2007. Chest wall thickness in military personnel: implications for needle thoracentesis in tension pneumothorax. *Mil Med* 172: 1260–1263.

Hoffman, H., Torres, W. E., and Ernst, R. D. 2002. Paleoradiology: advanced CT in the evaluation of nine Egyptian mummies. *Radiographics* 22: 377–385.

Knight, B. 1996. *Forensic pathology,* New York: Oxford University Press.

Luck, R. P., Haines, C., and Mull, C. C. 2009. Intraosseous access. *J Emerg Med* [Epub].

Recheis, W., Weber, G. W., Schafer, K. et al. 1999. Virtual reality and anthropology. *Eur J Radiol* 31: 88–96.

Verhoff, M. A., Ramsthaler, F., Krahahn, J. et al. 2008. Digital forensic osteology—possibilities in cooperation with the Virtopsy project. *Forensic Sci Int* 174: 152–156.

INDEX

T - #0641 - 071024 - C282 - 279/216/16 - PB - 9780367577025 - Gloss Lamination